POPULATION ECOLOGY

The Science and Society Series

This series of books, under the general editorship of the Biological Sciences Curriculum Study (BSCS), is for people who have both an interest in biology and a continuing desire to teach themselves. The books are short, highly readable, and nontechnical; each is written by a specialist on a particular, significant aspect of biology. Consistent with a long-standing premise of BSCS, the authors approach their topics as a base for inquiry; they pose questions and help the reader to see the way a scientist arrives at answers or where answers are to be sought. While the books are oriented for the layman primarily, it is hoped that they will also prove useful to high school and college students who wish more background in unfamiliar fields.

There are two major types of books in the series. One of these is directed to the continued and practical problems of society for which biology has both information and a message; topics such as birth control, eugenics, and drugs are of personal interest and concern to most individuals. The other cluster of books is devoted to the continuing problems of biology; these topics, such as animal behavior, population biology, and chemical coordination, often relate to societal problems and are of particular concern to biology as a developing discipline. Together, the two clusters give some measure of the totality of modern biology, to which we invite your time and attention.

William V. Mayer
Director, BSCS
Boulder, Colorado

POPULATION ECOLOGY

✳✳✳

Horace F. Quick

Biological Sciences Curriculum Study Book

Pegasus

A Division of The Bobbs-Merrill Company, Inc.

INDIANAPOLIS • NEW YORK

To Tim K. Kelley,
Geographer

Copyright © 1974 by The Regents of the University of Colorado
Printed in the United States of America
First printing

Library of Congress Cataloging in Publication Data
Quick, Horace Floyd   1915–
Population ecology.
(The science and society series)
"Biological Sciences Curriculum Study book."
Bibliography: p.
1. Animal populations. 2. Plant populations.
3. Ecology. 4. Human ecology. I. Title.
QL752.Q52   574.5'24   73–19664
ISBN 0–672–53678–1
ISBN 0–672–63678–6 (pbk.)

Contents

Editor's Preface

Boom! Bomb! Explosion! These are the epithets used by contemporary crusaders concerned about current population growth. Famine! Poverty! Environmental degradation! These are the consequences these crusaders associate with those epithets. That the world's population is booming/bombing/exploding is universally recognized and acknowledged. That the consequences may be other than famine/poverty/environmental degradation is not universally recognized nor acknowledged, and in many circles the likely consequences are hotly debated and/or denied. Here, then, is the kind of social issue to which this series of books is addressed, and here in this book is an examination of the way in which the biological sciences bear on the issue.

Earlier volumes in this series—Hardin's *Birth Control* and Volpe's *Human Heredity and Birth Defects*—speak to other aspects of this wide-dimensioned social problem. Professor Quick's book speaks directly to the biological, or more precisely, ecological principles which operate in determining the course of a population's growth and development. How fast, or slow, does a population grow? What natural limits does nature impose on that rate of growth? What natural limits does nature impose on the magnitude of that growth? What lessons on population growth and regulation do naturally occurring, nonhuman populations of plants and animals have for naturally occurring human populations? What lessons on population growth and regulation do various human populations have for yet other naturally occurring populations?

These are the kinds of questions which are explored in this book. To Professor Quick's credit, the exploration and explanation is accomplished in easily comprehended language and style without sacrificing the subtleties and sophistication of current theory and thinking in the field of population biology. Further, and most importantly, this is not a horror story on human population growth; the marketplace has been glutted with such books, and their function of arousing a public has been well served. This book is essentially a dispassionate discourse by a scientist, but it is not without feeling or concern about the central issue. After all, scientists are human beings, too.

Edward J. Kormondy
Olympia, Washington

Preface

Populations have one common cause: to survive. The survival of any population depends, in the first place, upon a favorable environment. This book is about the mechanism of population survival. It is not all about populations, because no one knows all about populations; but it is something about the intriguing events that seem to befall all kinds of life in what we might call their life histories. Living things are bound together in an environment that extends beyond the Earth, dependent upon the heat energy of the sun. The system is so intricately complicated and dynamic that all of the functions will never be known thoroughly. This is ecology. Ecology is about living plants and animals, including man, in environments that are suitable for life. The word is derived from two Greek words meaning "science of the house" and is also related to the word "economy."

In this decade of the seventies, a great concern for the welfare of mankind has developed. Because of the increasing social problems caused by human population pressure, it is more important than ever before for people to know about the rules of nature governing human events. *Population*, from the Latin word for people, is used to denote masses of humanity. Masses of people have certain fixed and certain variable characteristics. The analysis of these characteristics is called *demography*, a word formed from two Greek words meaning people and picture—a picture of the numerical and structural characteristics of masses of people.

Strictly speaking, *population* refers to people, but it has been generalized to apply to aggregations of all living things. It even has a mathematical use in the statistical sense, such as the number of things in a group called a "universe," or "population," by statisticians. The word is used in everyday conversation, however, to refer to human more than to other animal populations. We often read about insect, deer, or fish populations. Plants, also, are endowed with the attributes and characteristics of population dynamics, but people usually do not think of trees or grass as populations.

That mankind came into being on the planet Earth is without doubt the most important thing ever to happen. However egocentric this might seem, to contemplate man without man is impossible. Without man there is no meaning. Likewise, where there are environments without life, there is no ecology, because ecology implies the existence of living things in an environment. Ecology is the involvement of *the two components* of a natural system: *populations* of living things in favorable *environments*.

Environment can exist without man, but man cannot exist without an environment. In the title of this book, then, the word *population* is somewhat superfluous, since the term *ecology* includes populations. But we need terms that we can use to begin a discussion, and after they have served this use they may be revised or discarded, provided we do not sacrifice logic in the process.

So, we might properly use the word *ecology* while thinking of populations of trees, people, or geranium plants *and* the environments in which they flourish, but it would be incorrect to think of the ecology of any form of life without considering its environment. It follows that to think of the ecology of any one living species—deer, spruce trees, or people—is quite wrong, because ecology deals with the intricate interrelationships of all living things. Even so, the scientist usually must make some particular living thing, such as a wolf or a pine tree, the star of his show and view all other things that enter into the drama of the wolf or pine tree as supporting players.

All life is interrelated, interacting with the environment in a natural system of feedback loops and webs called an *ecosystem*. An ecosystem is something like a hot-water system or an air-conditioning system. It has parts, links, webs, flows, and stages. Without a population, there can be no ecosystem. Any environment without a population is simply a barren environment, like the moon or an empty refrigerator. On the other hand, a population without an environment is an unthinkable thing! But, a population in a favorable environment is an ecosystem and can be studied as the science of ecology.

Taking all these points into consideration, this book might better be titled *Populations in Their Environments,* but such a title would imply a content so comprehensive as to be unmanageable. Nevertheless, the aim of this book is to consider the population phenomena of some of the better-known plants and animals as a basis for a better understanding of human population ecology. The goal is to develop a perspective on principles of demography and ecology as a foundation for the political concept of *optimum population.*

POPULATION
ECOLOGY

I

The Geography of Primitive Man: Basis of Human Ecology

The concerns of modern man originated millions of years ago when human populations were distributed in environmentally controlled patterns. But populations were sparse then, and human beings were quite unaware of their impact on their own environment and of the beginnings of the population problems we know today. Gradually, as populations grew, struggles over a place to live grew into simple political pressures, such as tribal wars. Human ecology has evolved from primitive patterns of environmental relationships to very complex social and economic ones as well. The geography of primitive man is a key to human ecology. There was a time, less than ten thousand years ago, when different kinds of men were part of distinctly

different environments, and there were places where
there were no men at all.

A map showing the present-day distribution of the
human race reveals two simple but distinct ecological
relationships that have been the key to the evolution
of all life (Figure 1). Even today, mankind is con-

WORLD POPULATION DENSITY

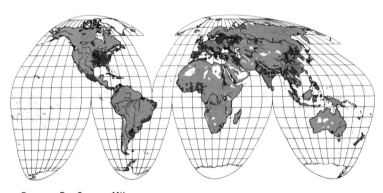

Persons Per Square Mile

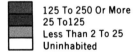

125 To 250 Or More
25 To125
Less Than 2 To 25
Uninhabited

Figure 1. The largest and densest populations are found in
temperate or subtropical regions that have optimum temperature
conditions for human beings.

centrated in the tropical regions of Earth, and the
polar regions are essentially uninhabited. From this
observation we might conclude that the tropics are
favorable environments and the polar regions are unfavorable. Of course, the in-between environments
could be expected to have favored the evolution of

different kinds of populations. And it is still a fact, in a broad way of speaking, that distinctly different races of human beings, having developed under diverse environmental conditions, dominate distinctly different parts of the world, even after centuries of migratory mixing. After all these years of acculturation, the basic ecology of man has not changed; he is still a physical being with a particular set of physiological requirements that can be obtained only from a favorable environment.

Prehistoric Distribution of People and the Provinces of Humanity

From the earliest beginnings of man, the distribution of people seems to have been quite definitely determined by geographic factors. In the first place, oceans form the greatest barriers; thus they are also definite boundaries. Terrain in general has shaped the process of human dispersal and migrations, particularly as seen in settlement patterns along coasts of seas, lakes, and along rivers. Mountains, also, have functioned in the separation of groups of people and in diversifying cultures.

The concentration of human population is, in itself, an indication of the character of places. We might assume that places where populations are most dense are the nicest places to live. While that is not always the case, the general idea seems to hold some truth. Populations are sparse around deserts or ice caps, such as the Sahara or the Antarctic. The dense population of India, on the other hand, would indicate that India is more amenable to human life than is the ice cap of the Antarctic. This can be said even though

India might not be the best place in all the world for the good life now.

Like those of today, prehistoric provinces had natural geographic boundaries, the most obvious being the great oceans and the next most obvious being the polar ice caps. The Americas, bounded at both ends by ice caps and on two sides by oceans, were a distinct province of humanity previous to the Spanish conquest barely five hundred years ago. Most of this vast landscape was inhabited by people with physical characteristics similar to the Mongoloids of Asia, but still recognizably different. Columbus called them Indians!

Africa, south of the Sahara, appears to have been populated by a distinct branch of the family of man, isolated by the great Sahara wasteland. Although the Nile River became a corridor for an exchange of population, the dominant race south of the Sahara has remained the same for thousands of years.

A vast province of arid and semiarid lands around the Mediterranean Sea—eastward through the Arabian and Syrian deserts to the barrier of the Hindu Kush and Himalayas—prehistorically was inhabited by people with common characteristics different from those of the surrounding provinces. Since the boundaries of one province also form the boundaries of adjacent ones, the Sahara on the South, the Alps on the northwest, and the Himalayas on the northeast encompass this area, sometimes referred to as the *Cradle of Civilization.* Mountain barriers to the north and east of that area form the southern boundary of the so-called Orient, dominantly inhabited—even to this day—by "Orientals." The island provinces of Australia and New Zealand were quite sparsely populated with people differing markedly from the Orientals and Caucasians of the Cradle of Civilization.

On the world physical map, a Eurasian province can be discerned as a rather vast lowland area with boundaries in common with the Oriental Province on one side, the Cradle of Civilization on another, and a great ocean and an ice cap on the remaining two sides. This is the province of the "European," a race of people who spread across formidable natural boundaries with such vigor as to displace the natives of three continents. Of all the races of man, this one has been the most restless (Figure 2).

Climate, also, has been an important factor in popu-

PRIMITIVE PROVINCES OF HUMANITY

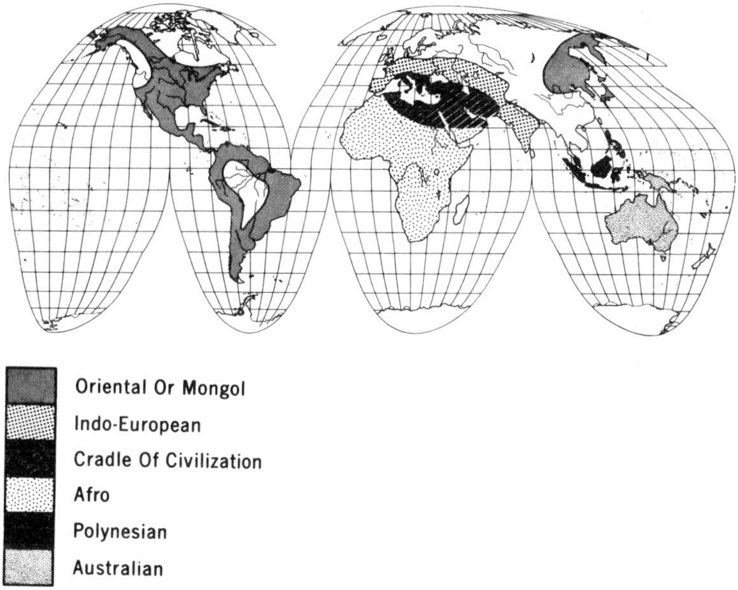

Oriental Or Mongol
Indo-European
Cradle Of Civilization
Afro
Polynesian
Australian

Figure 2. Anthropologists recognize certain Provinces of Humanity that are populated by races of people with differing physical characteristics. Even though migration has resulted in some mixing of the populations, the dominant racial distribution pattern has been the same for thousands of years.

lation distribution. Markham's study of population in relation to temperature has shown that the centers of early civilizations had average annual temperatures of 70°F., close to that of present temperate zones. Present-day population centers have a mean temperature of 61.1°F., nine degrees cooler than that of the ancient population centers. The historian Toynbee believes that this shift to cooler zones is related to the discovery and control of fire and also to the cultural improvement of shelter and clothing, events which have enabled people to move into suboptimum environments.

The body heat of human beings and most warm-blooded animals is within a range of two or three degrees. For example, the normal body temperature of man is about 98.6°F. The dog and horse have temperatures of about 101°F. This body heat is maintained as stable as possible by metabolic activity, but environmental conditions are not always favorable for doing so. In hot deserts, for example, a person can suffer heat stroke. If the desert air is hotter than normal body temperature and if the body can not dissipate or lose this heat, it overheats, causing death. In cold deserts or in polar regions, the body must have a great deal of food high in fat and carbohydrate to produce metabolic heat; it must also have heavy protective clothing to conserve that heat.

The rules of environment and those of physics and chemistry are in agreement—otherwise they could not function as "laws" of nature. One such law states, "Whenever and wherever conditions approach a minimum, the biota becomes impoverished"; in other words, populations of living things are denser where environmental conditions are most favorable. Where conditions are completely unfavorable, populations do not exist.

ENVIRONMENT CONTROLS POPULATION DENSITY AND DISTRIBUTION

RESPONSE OF INDIVIDUAL	Death	Torpor (Hibernation)	Active	Torpor (Aestivation)	Death
ENVIRONMENTAL HEAT	Too Cold	Cold	Moderate	Hot	Too Hot
POPULATION DENSITY	Absent	Sparse	Dense	Sparse	Absent

Figure 3. The basic pattern of human population distribution is controlled by optimum environment. Inasmuch as populations are composed of individuals, if environmental conditions are not favorable for individuals, then populations are sparse or absent. On the other hand, favorable environments have dense populations. Matching human body temperature with environmental temperature reveals that environment controls population distribution.

The idea expressed in Figure 3 begins with the assumption that temperature, or heat, is the basic environmental factor controlling life. Tests made with many kinds of animals have shown that the reactions of individuals to a whole range of environmental temperatures is basically the same. When it is much too hot or much too cold, organisms die. If temperatures are not actually lethal, some animals go into a stage of dormancy that is called aestivation under hot conditions and hibernation under cold conditions. In both cases, a lethargy overcomes the body and the bodily functions decrease. The hibernation of bears is well known, as are the summer sleep periods of many kinds of ground squirrels. The metabolic processes in both cases are reduced so that food intake ceases, respiration and heartbeats are reduced in frequency, and body temperatures are actually reduced.

Extremes of heat or cold cause death, of course, thus

eliminating the individual in such environments. A population is composed of individuals, and the individual that is not able to function or live in the environment dies; if all individuals die, then clearly no population exists! Thus, it is quite easy to see that distribution boundaries are environmentally controlled.

Even though allowances must be made for cultural adaptations that have enabled man to spread out and to multiply over the Earth, the simple test of whether conditions are favorable or unfavorable shows that there is still a basic environmental control of the density of human population, as well as a distribution pattern of the population that is in accord with the rules of ecology. India, with about 600 million people, has an average annual temperature of about 79°F. If this large population is excluded from the world average, then the present centers of densest population are found in regions with mean annual temperatures of 57.8°F. This pattern is attributed to the development of culture or technology, which has enabled people to occupy otherwise suboptimum temperature zones. Toynbee's viewpoint is that man needs stimulation in order to be inventive; world population growth has produced population pressure, which has forced man into suboptimum environments and thereby aided in the stimulation of more cultural or technical inventions in agriculture and industry. Thus, technology has enabled populations to flourish in new and sometimes less than optimum regions.

Other factors, too, have accounted for migration and the subsequent habitation of suboptimum environments. For example, the present centers of dense human population in the Americas started because of "push" factors. That is, people were pushed out of Europe because of political and religious oppression; they migrated to parts of the world which then were sparsely occupied wilderness. Later, better economic conditions in the New World resulted in "pull" factors that enticed people away from the older and denser population centers of Europe.

II

The Ecosystem Idea

Different kinds of plants and animals require different sets of conditions and materials to go on living. The sum total of these conditions and materials is called environment. The sets change from time to time and vary from place to place, and it is this change that has brought about the evolution of more than a million kinds of living things on Earth today.

A place where a species flourishes can be assumed to provide all of the conditions and materials needed by that species, whether it is plant or animal. Such a place would be an optimum environment. Further, since we can observe that some plants and animals grow together as communities, we can assume that the requirements of each form of life are provided by the particular environment in which they are found.

The concept of optimum environment is basic to the

ecosystem idea. It can be induced from observations of the presence or absence of different kinds of plants or animals in different kinds of environments. In this fashion, for example, the optimum environment for polar bears is found to be water, not land; salt water, not fresh; cold water, not warm. By matching factors in this fashion, we can describe particular characteristics of the environment where polar bears fare best.

A traveler going from New Orleans, on the Gulf of Mexico, to the Arctic Circle, along the 90th meridian, would see five major vegetation zones (black line in Figure 4). No matter what time of year he might make such a trip, he would notice a great change in temperature. The correlation of different kinds of plants with different temperature zones could be regarded as a cause-effect relationship. Scientists, generally preferring to test such observations, have conducted experiments in which different kinds of plants were grown at different temperatures, with numerous combinations employed. These experiments have resulted in the formulation of the "cardinal" temperature rule: each different kind of plant has its own upper and lower temperature-tolerance limits. If temperatures are too hot or too cold for a particular kind of plant, it will die. The survival of every living plant is highly dependent on a favorable physical environment (Figure 5).

If the same traveler were to become more curious about his observations and make a trip from Philadelphia along the 40th parallel of latitude to the Pacific Ocean on the California coast, he would see six major kinds of vegetation zones (dashed line in Figure 4). If he were to research their basic environmental characteristics, he would find that they all have different amounts of rainfall, and thus he probably would assume

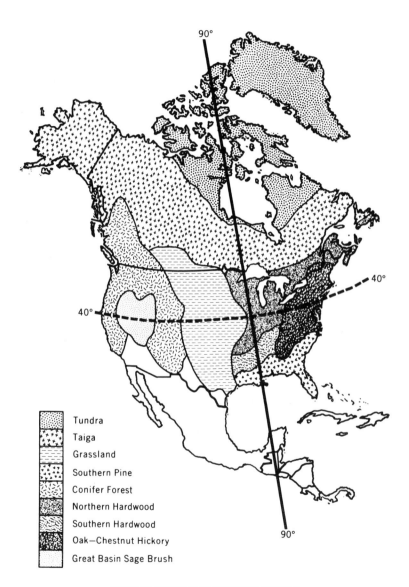

NATURAL VEGETATION OF NORTH AMERICA

that a particular amount of precipitation is one of the optimum environmental characteristics for each plant community. Again, tests have been made with controlled amounts of water given to groups of different kinds of plants. In fact, almost everyone at one time or other has conducted this experiment with potted plants at home. If we forget to water the plants, they die!

In this fashion, two components of environment—temperature and precipitation—are identified as basic physical factors of an optimum environment for different kinds of trees and plants, and each kind of tree or plant is a member of a population capable of surviving best only in an optimum environment.

Observations of this kind lead to the correlation of observed facts, and then to the generalization of theories. For the purpose of understanding population ecology, the environment must be taken apart, so to speak, and the various factors studied to learn their effects on living things. The role of temperature seems to us, at our present stage of knowledge, all-important. Without warmth there would be no life. However, extremes of temperature, ranging from the sun, which is too hot for life, and the moon, which is too cold for life, exist where no life can.

Environment has played the major role in populating the Earth with its present complement of plants and animals. Populations of plants and animals have flourished and failed time and again in particular

Figure 4. The natural vegetation of North America differs from region to region according to environmental conditions, each region having an optimum environment for a particular dominant plant type. The plant types shown along the solid line respond primarily to temperature differences from south to north. Along the dashed line, running east to west, differences are caused partly by temperature and partly by rainfall.

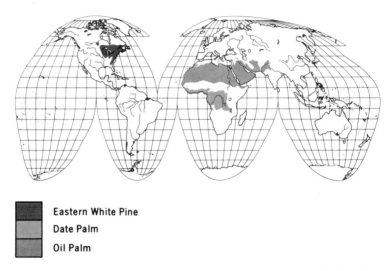

Eastern White Pine
Date Palm
Oil Palm

Figure 5. The natural distribution of three different kinds of trees is controlled by three different sets of favorable environmental conditions. There are thousands of such examples of how particular kinds of plants and animals are matched with differing environmental conditions in different parts of the world.

geographic locations. In those living things that have changed geographic range but have not changed significantly in form, their environmental requirements seem to have remained nearly the same. For example, horses that about 10,000 years ago were native in the area that is now known as Wyoming have since become extinct; but a very similar form has fared well in Europe and Asia up to modern times. About 250,000 years ago, rhinoceroses, similar to those currently in East Africa, inhabited parts of what is now called Nebraska. Thus we tend to believe that the environment of Nebraska once was quite like that of present-day Kenya and Tanzania. In this view, then, rhinos have changed their geographic range but not their ecological requirements. The change was probably accomplished by migration, a word that appears over and over again in the literature of evolution. In this example it gives the mistaken impression that a rhino packed up and moved to another country, when what must have occurred was a long succession of rhino generations drifting away from the changing environmental conditions unfavorable for the species. If they could not have moved fast enough, they would have become extinct. That is what is thought to have happened to dinosaurs. The giant lizards of Borneo, which are about six feet long, are probably descendants of the dinosaurs; they live in places far removed . from where dinosaur skeletons have been found but where environmental conditions are similar to those favored by their probable ancestors. Dinosaurs are not necessarily extinct, then, but have disappeared in form. Their progeny live as new forms, in new places geographically; but perhaps the places are "old" in the ecological sense.

Environments change on two main scales, long-term and short-term, and with these changes there occur, also, changes in the kind and number of plants and animals. Geological, or long-term, changes are manifest in such occurrences as the shift of the Earth on its axis, resulting in a shift of heat belts; this, in turn, causes a redistribution of life. There are hundreds of examples: the coal fields of the ice-covered island of Spitzbergen show that at one time the climate there was favorable for plants that formed the coal. Many giant buffalo skulls have been dredged out of the rivers of Alaska, but only the prairies have been their optimum habitat for the past ten thousand years. Obviously, at one time, climate favored the presence of prairies in Alaska, since the fossilized teeth of these giant prehistoric buffaloes show that they, like modern buffaloes, also were grass-eaters.

Short-term climatic shifts—for example, a yearly succession of wet or dry years—favor or disfavor plant and animal species and also induce variations in population size. When times are good, a population fares well, enjoying a high survival rate. Really tough times cause a decline in population size sometimes to a dangerously low point, but not quite to the point of extinction. Times of stress are common for all wild things and seem to occur periodically; such stresses produce a periodic, or rhythmic, change—called cycles—in population size.

It seems from the numerous researches of many investigators that the initial cause of variations in population size is a variation in the basic environmental factor: climate. Short-term variations are usually called weather. In the vicinity of Boulder, Colorado, for example, we witnessed the wettest summer on record in 1965; this was followed by the driest July on record in 1966. The unusually wet year, 1965, was a "mosquito year" in this otherwise semiarid region. The "mosquito year" was also a "sleeping sickness year," and all horse owners, of whom there are a multitude in the Boulder area, were quite concerned for the health of their horses. Further, the summer of 1965 was a busy one for local veterinarians.

The dry summer of 1966 was a "grasshopper year." Hordes of grasshoppers attacked the alfalfa and corn fields, and farmers had to call on the crop dusters for help. During this drouth things were tough for mosquitos, but somewhere a reservoir population of them survived. That was fortunate for the bats and swallows, which depend on insects as a source of food. This chain-reaction effect is an example of the way an ecological system works, but of course there are many more parts to it.

Such "thick and thin" cycles pervade the ecology of all living things, but they are more apparent among some populations than others. Even among such plants as the desert annuals—plants that grow from seed and live only one year—show a great response to short-range variations in weather. If there is no rain in late fall or early winter, seeds of the annuals will not germinate; a whole year of flowering might then be skipped. Perennials, such as trees, suffer only indirectly through the process of good and bad seed years, while the forest itself obviously "weathers through" variations within climatic patterns. In principle, then, we can observe that the length of population cycles is tied closely to reproductive capacity and environmental conditions.

Familiar climatic patterns composed of solar energy, precipitation, humidity, air pressure, air temperature, and wind—as well as less-familiar but similiar patterns called micro-environmental factors—affect the population dynamics of every living thing. These have "dynamic" properties; that is, they are always changing, and a change in one usually may be correlated with a change in another, such as a difference of altitude resulting in a change of air pressure.

The most familiar dynamic environmental factor is season. Spring and fall are ecologically expressive seasons; spring is the springing up of green grass after the dead of winter; and fall is the period of leaf-fall, when day length decreases after the autumn equinox and plant food production stops. If you look at a good atlas, you will see many maps that illustrate the dynamics of these basic environmental factors. Seasonal variations in different geographic locations are shown. For example, maps of frost-free periods show the number of expected days that will permit plant growth in a given locality. Storm tracks, seasonality and the amount of rainfall, the number of hours of daylight, and even the expected hours of cloudiness are environmental factors that influence both the distribution and the vigor of all the hundreds of thousands of living things on the planet Earth.

A number of discrete yet interacting components form our very complex physical environment, where some 300,000 kinds of plants, 4,000 kinds of mammals, 180,000 kinds of birds, half a million kinds of insects, and a myriad of such lower forms of life as yeast, bacteria, sponges, corals, and oysters survive. An ecosystem consists of such interacting components. The idea of ecosystems recognizes that there is an interrelationship of a number of subsystems, such as the hydrologic cycle, the atmospheric cycle, and the nutrient and gas cycles, which are absolutely necessary to sustain life.

Even though the living members of different ecosystems differ from place to place, there are common principles that apply to all ecosystems. A principal characteristic of all ecosystems is the food-chain relationship by which energy moves through the system. Energy-exchange levels are referred to in generalized terms as (1) producers, those that convert solar energy to plant tissue (grass is an example); (2) primary consumers, such as mice or cows, that eat grass; (3) secondary consumers, such as foxes, that eat mice; (4) sometimes tertiary consumers, such as some fish; and (5) decomposers, such as worms and bacteria, that break down leaves and animal bodies into the chemicals that were incorporated into the food chain by the primary producers in the first place.

The linking of these energy levels is sometimes called a food chain, or food web; it involves another idea of ecology, a pyramid of numbers—namely that there are more individuals in one energy-exchange level than there are in the next higher one. Now the complexity of population ecology can be seen; not only is there a pyramid of numbers, but each level in the pyramid consists of populations, all of which are subject to the laws of environment and function in accord with the principles of population dynamics.

The idea of the pyramid of numbers, or food chain, is really quite basic to population ecology. It is a phenomenon that influences even human ecology. Most fish that we eat, for example, are carnivorous and depend on a population of smaller fish, which in turn depend on a population of small floating plants called phytoplankton.

A realistic idea of the pyramid of numbers can be observed at a ring hunt in the African savannas. The ring hunt is a very ancient hunting system. Preparations for the hunt are made by drum signals. For several days the scattered villagers communicate by drumming. The message is passed along from village to village, and when it has reached all the bomas, or villages, of the region, people from each boma begin walking across the savannas toward a designated place. The circle of drivers begins to converge from points eight to ten miles apart. Drummers keep other parties informed of the progress of the hunt, and gradually the various parties close in. Of course, many animals escape through the ring. In fact, more escape than are caught and killed, but finally, as the hunters tighten the circle, they club and spear the animals. At the end of the hunt, the animals are counted according to kind and apportioned out to the hunters. There are always more small animals like duikers than large ones like waterbuck. The following list of animals killed during such a ring hunt in Uganda was recorded by an African game guard:

2 waterbuck	400 lbs. each
6 kongoni	300 lbs. each
14 kob	150 lbs. each
21 bushbuck	60 lbs. each
32 duiker	15 lbs. each

This list typifies the pyramid of numbers. It shows that an ecosystem consists of many more small animals than large ones. It also relates to the spatial distribution of population units, such as family and herd units, the herd units of smaller animals requiring less space than those of larger ones.

Of course, thousands of cane rats, hares, and rabbits escaped through the ring of beaters, which is to say the pyramid of numbers is much more complex than the results of the hunt indicate. Further, the animals listed are all herbivores, primary consumers; the hunt record gives no accounting of the carnivorous mammals, such as mongooses, genets, civets, leopards, and lions. Even though the numbers of the animals killed is only a sample of the whole animal population, it gives some real-life idea of the pyramid of numbers.

The results of a ring hunt may suggest that a vastly complicated and interrelated complex of animal and plant species exists within one definable ecosystem. The dynamic properties of species populations—their changes in number and their population characteristics, such as sex ratio, birth rate, and death rate—are so complicated that it is virtually impossible even for a team of scientists to unravel at one time all the mysteries of the population mechanism. But the system can be separated into component parts and the parts studied one by one. In that way, patterns and principles can be revealed and used to construct models of how the whole system works. For work it does, as is proved by its very existence over very, very long periods of time.

One way to view the functioning of some particular part of an ecosystem is to consider the role of a single member at the consumer level with respect to its "producer" base (which itself could be a primary consumer). For example, a lone fox is a secondary consumer, preying upon a population of small mammals that are primary consumers, or "converters" of plant material to animal flesh. A fox feeding upon a population of field mice is a common example (Figure 6). The

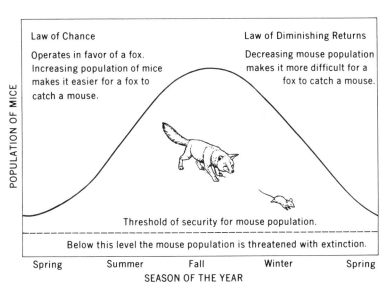

Figure 6. Predation is part of population ecology. A prey population varies from season to season in response to environmental factors, of which predation is a part. Predation of a single fox on a mouse population follows a pattern in which we can discern the operations of the law of chance and the law of diminishing returns. Ecology and economics thus are related in principle.

effect of the fox's hunting on the mouse population follows the law of chance and the law of diminishing returns. In the fall, when the mouse populations are highest, foxes find easy foraging. The law of chance functions in such a way that the more mice there are, the easier it is for a fox to catch one. This predation reduces the mouse population in number. The law of diminishing returns then begins to function as it becomes less and less profitable for a fox to hunt field mice. In times of great stress, foxes will abandon their predatory habits and change to a diet of vegetation. Even grizzly bears will eat grass when meat is scarce.

A fox's prey consists of a number of different animals, such as mice, squirrels, birds, and insects; the laws of chance and diminishing returns still apply, but no single species bears the whole brunt of predation. The most abundant suffers first, until its risk is equalized with another prey population; then these two species share the burden of predation until they are scarcer than some other kind of prey. As complex as this might seem, it is really quite oversimplified, for we have observed just one predator in this picture: a lone fox.

Often, in a given area, there is a complex of predatory animals as well as a complex of prey animals. A fox, like all predators, must share prey populations with competitors, such as weasels, owls, and hawks. A continual predatory pressure is exerted on prey populations. Owls hunt by night and hawks by day, some preying on the same species. There are bird hawks and mousers. Predation follows the law of chance when prey is abundant. However, the law of diminishing returns begins to work against all predators to favor the survival of prey animals when prey populations are scarce. Without such a survival mechanism, some species would become extinct. For that matter, sometimes they do become extinct in small areas, but this is a temporary population phenomenon if the environment is otherwise favorable. This illustrates the systems dynamics of an ecosystem at work, but only in part.

While there is much more to the population ecology of predators and prey, these examples show two underlying ecological characteristics: one is that each and every species of plant and animal responds to the basic principles of population dynamics. Another is that each species has its own spatial distribution characteristics. A fox, or a hawk, must be mobile in order to find food, and it moves over a more or less defined area in doing so. This dependence of the individual upon space for its food sets up a natural, competitive allocation of the environment. Ecologists call this *home range*. The word *territory*, on the other hand, has special meaning in animal ecology and relates to the spatial dimension of breeding activity. This is easily observed by watching a robin defend an invisible ring-like area around its nest from another robin. Encroachment of one individual into the territory of another usually results in a fight. The intrusion of a bull moose into the territory of another bull moose usually triggers a bull fight of the first order.

In other words, a territory lies within an individual's home range. Home ranges might be partly shared, but territories are not. In the wild, these conditions prevail between individuals of the same species, but frequently different species will appear to share the same space. In reality, they are sharing only the space, not the food. Even the gregarious wild herds of grazers, such as zebras, wildebeests, gazelles, and giraffes, which

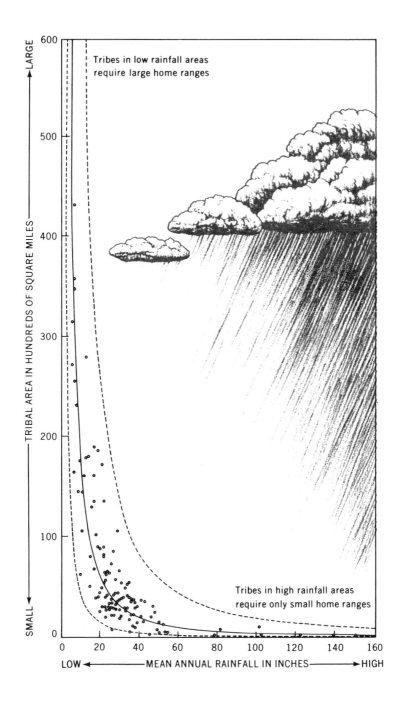

might appear to be feeding on the same stuff, are usually exploiting different plant populations.

It is easier to recognize tribal territories among primitive people than among people in more highly developed societies. Nevertheless, the same biological compulsion to define territory exists, even in modern societies. Primitive tribes are more visibly dependent upon the natural resources of their environment, however, and that usually results in a natural allocation of space to clans or other subgroups of tribes. Birdsell's study of Australian aborigines is a good example (Figure 7). Here, in a very arid land, water is the obvious limiting environmental factor. Springs or water holes serve as the center of a home range for a tribe or clan. As long as the human population is maintained at a level that does not exceed the water supply, the spatial pattern of population distribution can be maintained. Desert oases function this way, also. In ancient times, when population excess did occur, tribal war often ensued. Modern nations have established boundaries around their national estates or soverign areas; these resemble home ranges and reflect a natural, primitive urge to defend territory.

Countless times during the history of the human race, tribal wars to gain or protect territory have simulated animal patterns of competition for home range. Although environmental factors such as a limited water supply usually have triggered the conflict, population pressure is the basic cause of territorial strife. The rapid growth of a population in limited space is, without

Figure 7. In primitive societies dependence upon natural resources is more apparent than in developed societies. For example, water acts as a serious limiting factor to influence the distribution of Australian tribal groups. Where water is scarce, tribal areas are larger and populations more sparse.

doubt, the heart of animal—as well as human—ecological problems. Population growth causes environmental stress and environmental stress is the result of population pressure. This is how the cause and effect feed-back loop works in the ecology of all living things.

III

How Populations Grow

The oldest known living thing, a bristlecone pine tree in southern California, is about 4,100 years old. Most of us know that life on Earth existed thousands of years before this ancient tree was even a tiny seedling. We know, too, that nothing lives forever and that we are surrounded by living things that are just like their ancestors of 10,000 and more years ago; while things (domestic breeds of dogs and horses, for example) have changed quite considerably, wild things such as buffaloes and limber pine trees are just like the fossil remains of their ancestors that lived centuries ago. There have been, therefore, many generations of the familiar wild plants and animals that we see every day. Obviously, then, if all living things eventually die, and if their kind is present now, loss because of deaths must have been "evened up" by gain from births. The simple fact of the existence of any thing alive today proves that births have at least equaled deaths throughout the whole history of its kind.

The balance of births and deaths that enables a population of pine trees or elephants to continue living, even though all individuals eventually die, depends upon the principles of population growth form. Population growth form is a generalized term for the overall effect of ever-changing population factors, such as additions by births, subtractions by deaths, and migrations, that add to or subtract from the population of a given area.

Many children have kept tropical fish, especially the popular guppy. Guppies are prolific breeders, and in the first stages of population growth, births invariably exceed deaths. This is the most usual pattern of population increase and is called by biologists Malthusian, or exponential, growth. It is the most common growth form in the early stages of all population development.

Factors of Population Dynamics

The growth or decline of populations is a result of action by the factors of population dynamics: births, deaths, and migrations. Migrations can take place out of an area (emigration) and into an area (immigration). The effect of emigration numerically is like that of death for a particular area being left, but the person automatically becomes an immigrant in his new home, and the effect of immigration becomes numerically equivalent to births in the new home area.

Changes of size and structure resulting from births, deaths, and migrations easily can be observed in populations. In discussions of population dynamics, space requirements usually are ignored because it is simply too much to consider space, births, deaths, and migrations all at the same time. They cannot be ignored completely, however, and will be discussed later with the importance they deserve. The spatial aspect of environment involves matters that favor or disfavor a population and therefore affects its growth form. First, however, our discussion will center on the main ideas of population growth form that result from the interaction of birth, death, and migration.

Exponential Growth Form

Fruit flies, the little flies that seem to come with bananas, have also—like guppies—been the subjects of many experiments on population growth. These inexpensive organisms are easy to keep and quick to multiply. The experimental biologist can learn more about population dynamics in a shorter time from these organisms than from long-lived animals such as horses and human beings.

Experiments to test population growth are rather easy to do in the biological laboratory. The procedure involves putting a known number of living things in a favorable environment and then counting the population at given intervals of time. If the subjects reproduce sexually, then of course at least one male and one female must be put into the tank, cage, box, or whatever is used to provide their environmental requirements. The data can be recorded in a simple table with two main columns: one for time and the other for total population.

If a male and a female guppy were put into a suitable tank of water and given an appropriate amount of food, the number of guppies would increase naturally as a result of the inexorable law of self-perpetuation. If there were no environmental limitations on the growth rate of the population, periodic counts of the numbers of guppies would reveal an unimpeded increase. Table 1 shows a theoretical pattern in which the population increases with each unit of time.

TABLE 1
Exponential Growth Form

Time	Adult Population	×	Birth Rate	=	Increase of Young	+	Parent Population	=	Total Population
1	2	×	2	=	4	+	2	=	6
2	6	×	2	=	12	+	6	=	18
3	18	×	2	=	36	+	18	=	54
4	54	×	2	=	108	+	54	=	162

Although it may look like a stuffy table of numbers, Table 1 implies certain attributes—the ability to reproduce (birth rate), a gestation period (one time interval), and complete survival (no deaths)—of living things. It adds up to what biologists call "unimpeded increase." This means that deaths in the population have been ignored in calculation, impossible to do in actuality, of course. For our purposes here, however, such a device is allowable to help explain population dynamics. There are several other terms for this "unimpeded" growth form: exponential growth, Malthusian growth, or even geometric progression.

Logistic Growth Form

Many people have observed that living things generally produce seeds or young far in excess of the numbers that can reasonably survive. Darwin made such observations, and from them he developed some of his ideas about natural selection and survival of the fittest. Malthus, an English economist who lived at the time of the

American Revolution, noted that a population tends to increase faster than its environment can support it. Although the views of Malthus have not proved entirely correct, the principle that populations increase geometrically while food increases only arithmetically is demonstrated frequently by the multiplication of wild animals during periods of food shortage or severe weather.

Malthus's idea can be tested by continuing the exponential growth form experiment on guppies or fruit flies. The usual result of such an experiment is that the population increases rapidly (unimpeded) at first, and then gradually slower. This pattern, called *logistic* growth, is so fundamental in biology that even the growth pattern of an individual follows the logistic curve (Figure 8).

Experiments have revealed that something in the environment will not let population growth continue at exponential rates. Many observers have assumed that some element in the environment is not available in sufficient quantity to permit all of the animals to survive. Malthus, himself, suspected that a food shortage leading to starvation would be the first factor that would check the growth of the human population and thus prevent it from following the exponential form of population growth.

Feeding experiments beginning with very young birds or mammals show that a weight gain day by day follows the same pattern as population growth. If the daily weight of a puppy is plotted over time on graph paper, an S-shaped curve, the same as the logistic curve, will be formed by a line connecting the points. Many animals, especially warmblooded ones, do not grow any more after maturing. This appears to be inherently, or genetically, controlled. The weight or height growth

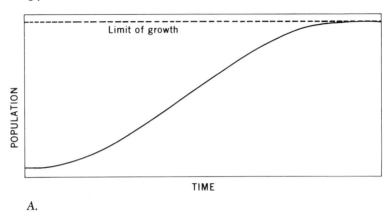

A.

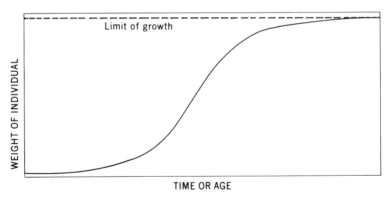

B.

Figure 8. Logistic growth form is so common in nature that both individual growth and population growth follow the same basic pattern, sometimes called sigmoid because of the S shape of most growth curves in nature. A. Size of population increases exponentially as time goes on until the limit of carrying capacity is reached. B. Individuals of most animals grow rapidly during youth, but the rate of growth is reduced with age.

of children shows this same pattern as they grow older; growth ceases when they mature. Their growth curves level off at the top of the **S**. Cold-blooded animals such as fish and snakes, or lizards and crocodiles, continue to grow throughout their whole life time, but the rate at which they increase in length or weight is much less in old age. A plotted curve of their growth would also be S-shaped, but in old age the top part of the **S** would continue upward at a slowed rate rather than level off.

In the case of whole populations, the leveling off of population growth is assumed to be caused by some factor limiting this growth. The most common environmental factors that limit wild animal populations are food shortages which result in starvation or serving as a food source for some other animal. In human populations, Malthus described these factors as famine and war. If they do not occur, the next most probable control factor, according to Malthus, is disease. We now have evidence, however, that disease usually strikes down a population before starvation.

Disease, itself, is a population phenomenon: microscopic organisms, or "germs," that produce disease grow and develop when the human body serves as a favorable environment. Malaria is an example. When a population of malaria carriers, one-celled agents, reaches too great a concentration in the body, a fever results. White blood cells then increase in number and destroy the malaria parasites. This form of "predation" slows the growth of malarial population, and had we been keeping account, we would have found malarial growth form, thus exhibited, to be sigmoid. The same would be true of the white blood cells, but a lag would be found in their buildup; their increase would occur at a later time. Because the two are "out of phase," the increases and decreases of the populations in relation to each other would cause cyclic, or periodic, fever. These population "cycles" will be discussed more fully in a later section, but they serve here as an illustration of two logistic population curves that are dependent on each other. Logistic growth form is a basic feature of population ecology. It is a way of explaining and measuring the dependency of population dynamics on environmental factors.

Age Composition of Populations

General experience tells us that populations consist of individuals of different ages: the young, the adult, the elderly. Age composition, therefore, is a characteristic of populations. Table 2 is another kind of population-growth table, one that records changes in the structure of a population as time goes by. Let's assume that two animals were put on an island where none of their kind had ever lived before. For the sake of simplicity, we shall assume also that they reproduce asexually at a constant rate.

TABLE 2
Development of Age Structure as a Result of Growth Form (No. of Animals at Each Age Level)

Year	No. of Population in Each Age Class					Total Population
	1	2	3	4	5	
1	2					2
2	4	2				6
3	12	4	2			18
4	36	12	4	2		54
5	108	36	12	4	2	162

Perhaps the easiest thing to see in this model is that the two animals that began this population as one-year-olds in year one became five-year-olds at the end of five years (follow their "aging" diagonally from top-left to bottom-right in the table). By the end of year two, the population has developed two age classes (four one-year-olds and two two-year-olds). It has begun to show a definite structure. At year three, there are three age classes, and by this time the youngest age class is proportionally much larger than any other single age class. The ratio of young to the *total adult* population is constant (2 to 1), however, because we have assumed a constant reproduction rate of two young per adult each year.

It is easy to read from such a table that by the end of five years there would be five age classes, and that the population by this time would have a recognizable age structure. Children who keep pet rabbits or tropical fish are actually conducting population growth experiments, but they usually do not think of the process in terms of population dynamics. Natural populations also develop age structure as a consequence of growth form in the same manner that experimental populations do, and quite in keeping with the model shown in Table 2.

An age-structure curve can be plotted as in Figure 9B from the columns of age figures for year five. This graph slopes *down* to the right, but a growth curve plotted from the change in total population over the five-year time period would slope *up* to the right (Figure 9A). One is similar to a mirror image of the other, and a close look at them both shows how they are related. One of the main things to observe is that on the population growth curve, TIME is the scale along the bottom of the graph, but AGE is the scale in population

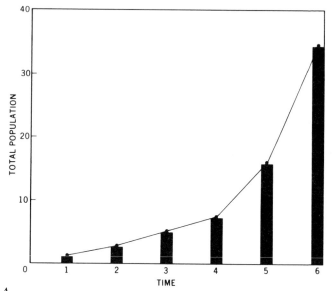

A.

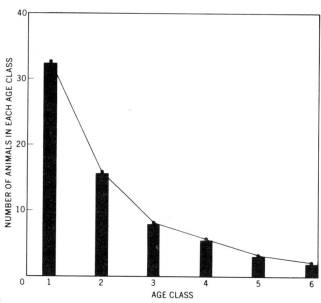

B.

structure. Time and Age, of course, are related. As we get older, time goes by; or as time goes by, we get older.

A more practical view must be taken now, in retrospect. In these mathematical models, not only has death rate, or mortality, been ignored, but the effects of sex and the age of reproductive maturity also have not been considered. If they were, however, the same general patterns and principles would appear. As we shall see later, there is usually a different rate of death for males and females, which changes the numbers but not the basic ideas of population growth form and the development of age structure.

The basic concept of population biology is growth form, both exponential and logistic. Growth form generates variations in the age and the sex composition of populations. The inherent reproductive attributes of a species guide the patterns of growth form to an extent, but the forces of environment are powerful determinants of survival and therefore of population size and composition. This is the arithmetic of population ecology in elementary form.

Figure 9. Age structure, or composition, develops naturally from normal population growth patterns. *A.* Typical growth form is "exponential," also called "geometric" and "Malthusian." *B.* As a population grows, young are added to the population and adults get older, resulting in several to many age classes.

IV

Demography, or How Human Population Dynamics Are Studied

People who study human population dynamics are called demographers. They recognize a distinction between pure demography and population analysis: whereas the study of demography usually is limited to quantitative changes in populations, such as size, sex ratio, or age structure, population analysis deals with social, economic, and political events that are related to these changes.

Although censuses are the basis of all demography, the most powerful tool of the demographer is the life table, a compilation of vital statistics from census data, or samples of living populations, or records of deaths. Life tables reveal numerical, or quantitative, descriptions of population dynamics, otherwise called vital statistics. Other tools that can be derived from the same data are age pyramids, survival curves, and mortality curves. All of these, however, give purely quantitative descriptions of many different characteristics of a population. In other words, the dynamics of the population can be expressed with these tools, but the social and economic conditions are hidden in the data.

At first glance it seems contradictory to use these devices of demographic analysis to explain the dynamics of a population at a fixed, or static, point in time. Despite the apparent contradiction, however, we can learn to read life tables, age pyramids, and mortality curves as if they are pictures of not only a particular moment in time but also what is going on in a population. Death rate and birth rate are examples: the word *rate* implies change with time, yet states a momentary condition.

Although life tables tell the story of structured changes in populations, they do not tell us anything about size or spatial distributions. Many economic and social phenomena show up in the life tables, of course, but usually we must have some facts other than the tables themselves before we can interpret how these phenomena have affected the data. For example, just after World War II the life table of the Russian population showed a very noticeable shortage of males between the ages of sixteen and forty. The high war casualties for a period of eight or ten years caused the shortage, but unless one knew this, the life table would show only that unspecified mortality occurred in certain age classes and mostly among males.

In the age pyramid of the United States (1970 decennial census), there is a notable constriction of the age classes between thirty and forty. Both males and females are underrepresented when compared to the usual curve of population growth form. This constriction reflects voluntary birth control during the depression of the 1930s. People who were thirty to forty years old in 1970 were born in the period from 1930 to 1940, the "baby depression" years. These events are part of the ecology of populations; they illustrate that the ecological history of a population is concealed in its life table.

Life Tables

In Chapter III we saw that there is a natural flowing relationship between population growth form and age structure, and that age structure is created by births, deaths, and migrations. Also, mortality appears to be the most obvious factor that generates a particular structure in a population. It is true, of course, that if there *were* only increases by births into a population, there would still be a particular structure to the population and each age class would continue always in the same proportion to the whole population. Although we must admit that birth and migration are factors producing changes in population age structure, mortality is the most effective determinant of a population structure and more particularly of the demographic differences between different nations.

Life tables are believed to have been devised in 1690 by Halley, the already-famous astronomer, who first observed what is now called Halley's comet. Before this, J. Graunt (1662) tallied up the "bills of mortality" from records, or "bills," of deaths caused by the infamous Black Death. This tabulation and others led to the idea of life tables. A life table is what your insurance agent uses to calculate his risk in insuring your life. Insurance companies have the same kind of interest in the pattern of human death as did Graunt and Halley, but for a different purpose. A life table is a kind of game "table," on which people bet against each other.

Two sources of data are available to the demographer or the actuary in calculating a life table: one is a tabulation of the number of deaths that have occurred in each age class, and the other is the number of people alive in each age class. Death data can be obtained from a registry of vital statistics at the county courthouse, but the number of living people in each age class must come from a census or a sample of the population.* These are the *primary sources* of such data.

Since it seems odd that they should be made from a record of deaths, life tables are sometimes referred to as *experience* tables. The demographer or actuary must know how much "life" their subjects have experienced before they can make a life table, for it is only upon death that the experience of life, or length of life, is finally measured.

Of the three types of life tables, the most commonly used is called *time-specific* and may be obtained from an "instantaneous" census of people alive, or from the ages at death during an "instant"—usually one year. Another type is called *dynamic*, or *cohort*, and relates to the population dynamics of a group of babies, cohorts, who begin life in the same year. This type of life table is rare because it can not be completed until the last survivor dies. However, we can use this type to study the population dynamics of short-lived species, such as

---

* The birth, death, marriage, and divorce of a person must be registered by law, usually with the Board of Health at a county seat. Demographers call these vital facts, used to calculate vital statistics.

wild ducks. Both of these types can be combined to make a third, the *composite* life table, particularly useful in wildlife studies because of the scarcity of time-specific or cohort data about wild animals.

In the simplified life table shown below (to see how calculations are made, see Appendix A), the numbers could represent people, trees, or wild goats. An array of data could be obtained from death certificates, such as in the "bills of mortality," and used in the middle column to represent deaths in each age class.

TABLE 3

A Simplified Life Table

Age Class	Number That Died in Age Class of 1000 Born	Number of Survivors at Beginning of Age Class Per 1000 Born
1	333	1000
2	267	667
3	201	400
4	133	199
5	66	66
6	0	0

At the beginning of age class 1 there were 1000 survivors—the number born into that age group. During that age interval, 333 individuals died, so the number of survivors at the beginning of the next age interval (age class 2) was 667 (1000 less 333), and so on. In age class 5, the remaining 66 individuals died, so there are 0 survivors at the beginning of age class 6. These "raw" data are used to calculate other components of the life table, such as probability of death and mean life expectancy, to be discussed later.

Probability of Death: A Vital Statistic

Perhaps the most important vital statistic to life insurance agencies is the mortality rate. This is also called "probability of death" and is the same as chance of death at a particular age, for which the demographer uses the symbol q_x in life tables. The following example, of course, is theoretical:

TABLE 4
Mortality Rate or Probability of Death

Age Class	Number That Died in Age Class of 1000 Born	Number of Survivors at Beginning of Age Class Per 1000 Born	q_x, or Mortality Rate ($\times 100$)
1	333	1000	$\dfrac{333}{1000} = 33\%$
2	267	667	$\dfrac{267}{667} = 40\%$
3	201	400	$\dfrac{201}{400} = 51\%$
4	133	199	$\dfrac{133}{199} = 67\%$
5	66	66	$\dfrac{66}{66} = 100\%$

To calculate the chance of death (mortality rate) for a given age class, we must assume that during that age period the number of deaths is a certain proportion of the number of survivors that begin the age period together. Therefore, the number who die *during* that same age period can be expressed as a fraction of the population entering that class, and that is the mortality rate—or chance of dying—at that age. It is like the gambler's idea of chance, namely, "so many chances out of a thousand of dying during this age period." Thus, for the first calculation (Table 4), the number of deaths (333) is taken as a proportion of the number of survivors at the beginning of that age class (1000) or $\frac{333}{1000} = 0.333$. Probabilities are typically expressed as a decimal, but for our purposes we have multiplied by 100 and expressed the mortality rate as a percentage. Thus, during age 1 to 2 there are 267 deaths out of 667 living things, or 400 per thousand, which is the same as forty percent chance of death during this age period.

Another Vital Statistic: Mean Expectation of Life

What is the average lifetime of an American or an Asian, a robin or a pine tree? Demographers have two terms that relate to this attribute of population dynamics. The most comonly used term is "mean expectation of life at birth," and this refers to newborn babies. When this term is applied to newborn infants it means they can expect to live so many years; this is the predicted average length of life for all of the babies born in the same year. Some will live longer, others will die sooner, but the average is expressed in a life table as the mean expectation of life at birth. Mean expectation of life explains a great deal about the ecology of a national group, an ethnic group, or a species of plant or animal. For example, mean expectation of life for a mouse is less than a year; for a sea gull, about twenty-eight months; and for an American, about seventy-two years.

There is also a second meaning of this vital statistic. Beyond the first year, each member of an age cohort who survives the first year has a new life expectancy, designated "mean expectation of *further* life." It is like running an obstacle course; if we get by one hazard, we get a chance to try the next! In some societies and in some animal populations, mean expectation of life is higher if one manages to survive infant mortality, but in general it is highest at birth and diminishes each year. This, of course, reflects the inevitability of death for all living things.

The life expectancy of the theoretical population in Table 4 is shown in the last column of Table 1B, Appendix B. If the age class is in years, the average individual at birth has a life expectancy of 1.83 years, and with each advancing year his life expectancy decreases. But, as pointed out above, just because the organism is a year older, his life expectancy doesn't necessarily decrease by an even year. The method of calculating life expectancy is shown in Appendix B.

Other Features of Life Tables

Longevity, another vital statistic shown in a life table, does not have to be calculated. It is shown directly as the oldest age class in the life table. In wild populations that are subject to the hazards of a wild existence, the longevity of a species of animal is called ecological longevity, because environmental hazards generally shorten their lives before they die of natural old age. Domestic animals such as dogs and cats, protected from natural environmental risks, may live to reach a physiological longevity and die of old age.

Birth rates can also be calculated from a life table, as they can from age pyramids or survival curves, but the procedure is not apparent from the table itself. The ability of a population to evade extinction—to maintain itself from generation to generation—can also be determined from data in a life table, but the standard table does not show this dynamic property of a population, either.

A life table is a way of keeping score of the population ecology of a particular living thing. The vital statistics are quantitative measures of a species' response to its environment. In chapters to follow, some of these relationships will be illustrated with life tables, age pyramids, and survival curves of different kinds of plants and animals so that we may compare the population dynamics of some other living things with those of human populations.

Demographers often make sex-specific life tables to show the differences in the vital statistics of men and women. Women, for example, usually live longer than men and therefore at birth have a greater mean life expectancy than men do. The general life table, on the other hand, lumps the census data together for both males and females. So the demographer has three life tables to use in analyzing population dynamics: the general life table, one for females, and one for males.

Age Pyramids

Age pyramids are quite popular ways to illustrate the age structure of the populations of different countries. Perhaps this is because the age pyramid shows the proportion of the sexes by age class, where two survival curves—one for males and one for females—would be required to show the same population structures. Making an age pyramid is somewhat like turning a survival bar graph, such as Figure 9(B), on its side and placing the bar that represents the youngest age group on the bottom. The difference, however, is that pyramids usually show each age class as a *percentage of the whole population* while most survival curves show each class as a proportion based on one thousand.

Conventionally, age pyramids are divided vertically; the bars representing males are shown to the left of center and those representing females to the right. The horizontal extension of the bar graph to each side is in proportion to the ratio of each age class by sex to the whole population. If all the bars are added together they would equal one hundred percent of the population. This shows again that, like the life table, the number of people, or the size of the population, is not represented at all, but only the structure of the population: proportion of each age class to the whole. Sometimes the age class intervals are shown in a column up the middle of the pyramid to separate the sexes more clearly* (Figure 10).

The worst problem that the demographer encounters in dealing with these graphic aids, and with the life tables, too, is that he is trying to discuss and consider dynamic properties such as birth *rates* as if they didn't change. To perceive changes in populations, he needs age pyramids or life tables pertaining to two different time periods. That is why censuses are taken periodically, usually every ten years, to check on population growth and changes in its structure.

---

* The scale at the bottom of an age pyramid should be carefully inspected, because several variations of scale are used.

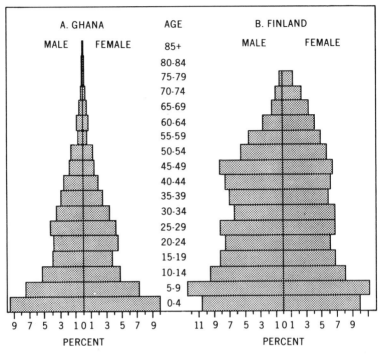

Figure 10. There are three basic types of age pyramids: constrictive, expansive, and stable. Each type reflects demographic, economic, and social characteristics of a national population. *A.* The age structure of Ghana is expansive. This reflects a high birth rate and a high ratio of persons who are dependent on the labor force. *B.* The Finnish population is nearly stable, with a low birth rate and a low dependency ratio. Consequently, literacy and health rates are high because the labor force can more easily afford a good school system.

Later, age pyramids from different countries will be compared to show certain economic and social conditions that prevail in different human populations. At this point some generalizations can be made to show how age pyramids can be used. An "expansive" age pyramid, with broad base and sharply sloping sides, shows a population, such as that of Ghana (Figure 10A), with a high birth rate and a high death rate. This population would have, among many other social and economic attributes, economic difficulty in supporting a school system for the young dependent group. A population pyramid with a narrow base, such as that representing Finland (Figure 10B) would indicate no such problems. The ratio of the young dependent group is low when compared to the productive segment of the population called the labor force; it could more easily support a good school system.

Survival Curves

Another demographic tool, the survival curve, is also a visual, or graphic, method of showing the fate of a population. Survival curves can be plotted from life tables or taken directly from a census. Such a graph would show generalized dynamics of a whole population, unless, as in age pyramids, the data were separated by sex in the beginning. Plotting a survival curve from a life table merely means making a pencil mark on graph paper for each age class opposite the number of survivors in that age class. If the plotted points are then connected, a curve is formed by the connecting line. As in Figure 11, it is customary to put population or number on the vertical axis and age class on the horizontal axis.

A survival curve can also be illustrated as a bar graph, called a histogram (meaning height drawing) (Figure 12). In most populations, each successive bar is shorter than the preceding one because of mortality, or loss, as the cohort progresses through time from one age group to the next older one.

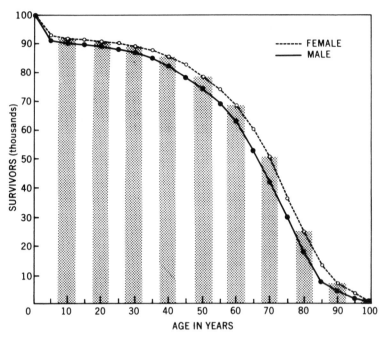

Survivorship curves for white males and females in
continental United States, 1929 to 1934.

Figure 11. Survival curves, like age pyramids, are of three
basic types: diagonal, positive rectangular, and negative rectangular. This survival curve of the population of the United
States during 1930 is of the positive rectangular type, indicating
a high rate of survival. Compare with Figure 12, the negative
rectangular type, which indicates lower survival rates. The
diagonal type (Figure 13) appears as a straight line from top-left
to bottom-right, indicating a constant mortality rate; the reciprocal is a constant survival rate between age classes.

Another way of showing survival is to make "shadow
stripes" between each survivor bar, indicating the numbers of the population who have died during the preceding time interval. A survival curve of the African
elephant (Figure 12) is an illustration of this. Thus, the

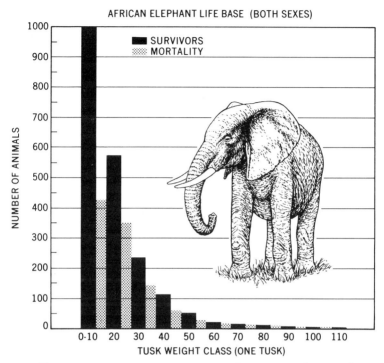

AFRICAN ELEPHANT LIFE BASE (BOTH SEXES)

Figure 12. In contrast to American human populations, the survival pattern of African elephants shows a high mortality rate among young up to about age fifty, at which time the hardened adults have a higher survival rate. The steep down-slope of the curve shows high mortality, which is the reciprocal of low survival. This is the negative rectangular type of survival curve that is characteristic of most wild animals.

number of survivors in tusk weight class 30 is equal to the number in tusk weight class 20 minus the "shadow stripes," or mortality, that intervenes between these two survivor classes.

Probably the most useful feature of survival curves is in estimating the fates of different populations. To illustrate this, two groups of curves are shown in Figure

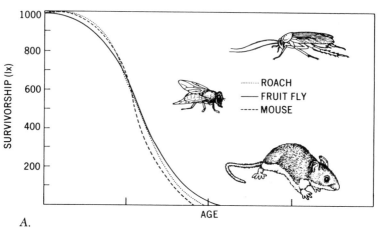

A.

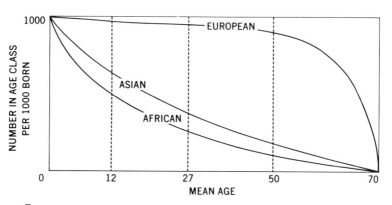

B.

Figure 13. A. Three different kinds of animals, all with similar survival patterns called diagonal. Most animals—insects to mammals—have high birth rates, high mortality rates, and short lives. B. Three different ethnic groups of human beings closely simulate the three basic types of survivorship. The European group is of the positive rectangular type; the Asian is of the diagonal type; and the African is of the negative rectangular type. Each group has a different life expectation at birth, the result of differences in mortality.

13. Part A is a comparison of three kinds of animals and Part B is a comparison of three ethnic groups of people. Most biological data indicate that there are three basic types of survival curves. Although these curves are used to reflect survival directly from age group to age group, they also indirectly reflect mortality, as if there were "shadow stripes" between each age group. A differential survival rate among different kinds of animals is well known, as shown in Part A. Part B illustrates this feature in demographic (human) terms. The upper curve, referring to "Western" types of populations, shows very high survival rates—the same thing as very low mortality. In contrast, the lower curve representing African populations shows very high mortality, even though birth rates are sky-high, indicating consequently, low survivorship.

Part B of this figure reveals that, almost always, certain political and economic conditions are dependent upon survival characteristics. For example, the curve illustrating survivorship in African populations shows a high ratio of dependents to productive labor force, which is relatively small. In such a population, the burden of educating the young would be very heavy.

Mortality Curves

There also are several basic types of curves or patterns which can be used to describe mortality in any population. Most trees, for example, produce an enormous quantity of seed. On the average, at least 95 percent of tree seed never even germinates. Mortality is high; fortunately, however, as time goes on those trees that do survive have an increasingly better chance of survival, until they approach their physiological longevity. Fol-

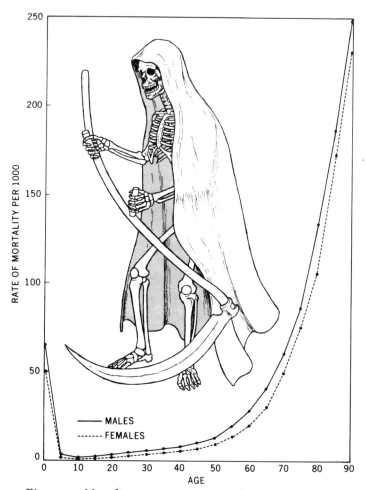

Figure 14. Mortality curves are commonly reciprocal in form to survival curves. This curve of the American population during 1930 shows about 50 deaths per thousand infants born. After infancy, the mortality rate is greatly reduced for ages 5 to 40, at which age mortality rates become increasingly higher. Note that at each age, the mortality rate shown is per thousand persons of that particular age. At age 100, death rates are nearly 1,000 per thousand, or nearly 100 percent, because this is practically the extreme longevity of the human species.

lowing the period of seedling mortality, the probability of death is less during young and mature stages; but as trees arrive at the height of their physiological longevity, the probability of death again becomes high.

Infant mortality in Scandinavian populations is very, very low, the lowest of all human populations. But there is a physiological limit to length of life, and as this is approached, the chance of death becomes more and more certain for the older age classes. The form of such a curve is like the scythe of the "grim reaper" (Figure 14). Other populations have mortality curves intermediate between those of plants and Scandinavians.

Although mortality curves are of considerable importance to the demographer in studying the causes and effects of death upon population structure, survival curves more directly reveal population ecology. Even more detailed knowledge of the ecology of population dynamics can be obtained from life tables. When the ecologist and the demographer combine their knowledge, the functions of population ecology can be more readily explained.

Other Vital Statistics

Demographers employ many kinds of vital statistics to study certain specific characteristics of populations. One of the most common is the "trend." Trends in population size, birth rate, death rate, and other vital statistics are comparisons of specific population characteristics over periods of time. The trend in population growth is most familiar and is determined by comparing the population of a particular country at one particular time to a different time.

One of the more popular measures of population

dynamics has been to estimate the "doubling time" of populations, or how long it will take to increase the present population to twice its present size. Concern about population problems has made this a popular conjecture for many years, but usually populations have not lived up to the demographer's expectations.

Nevertheless, there are many useful demographic statistics that help to explain social and economic conditions, but these are related more to cultural or ethical causes than to environmental ones. It might be said from the ecologist's point of view that the demographer has underplayed or even ignored the importance of natural environment in human affairs. On the other hand, environmentalists have tended to overlook the tools of the demographer and their usefulness in understanding cause and effect relationships in population ecology.

V

Familiar Population Dynamics

All Americans know the robin. It lives in every state of
the Union except Hawaii. It almost always lays four
eggs in the nest. It habitually pairs with a mate for the
breeding and nesting season. The male works as hard
to feed the young as does the female. They share the
work and fight for the survival of their species. Yet,
this two for one, or doubling of the population, never
seems to change the apparent number of robins from
year to year. When we consider their great geographi-
cal and ecological range, the vast total number of their
species, and the principle of the exponential growth
curve, it is a great wonder that we are not deluged with
robins. In fact, their mean life expectation upon hatching from the egg is only 0.7 years. Less than one year!
Inasmuch as robin nesting, or reproduction, is quite
definitely controlled by the physical environment, particularly light and heat, the species usually has only one
chance to nest each year. It would appear, therefore,

that if the average length of life is less than one year, many robins never get the chance to reproduce their kind.

What happens, then, to about half of all the robins born in one year? Foxes, hawks, weasels, magpies, and all the other small carnivorous creatures of the secondary consumer level depend, at least in part, upon this surplus of robins above their threshold of security for a livelihood. In turn, some environmental factor or group of factors tends to level all populations, including carnivorous consumers. Otherwise, we would be run off the face of the Earth by shrews, the smallest mammal in the world! Ecologists refer to this fluctuation of increase and decrease as *dynamic equilibrium* (Figure 15). This means that all populations vary dynamically above and below a mean, or average, population level. The mean is a measure of long-term equilibrium. We perceive that most species have existed for at least several thousands of years, and by observation we come to the conclusion that their populations vary from time to time.

Field mice and their aquatic cousins, muskrats, have the capability of producing about fifteen young per year. This usually occurs in three or four litters of about five young, born over a period of four or five months. Like robins, these consumers might also be expected to overrun the Earth in no time if there were no checks and balances in the ecosystem. The population dynamics of robins and field mice are vastly different in their physiological characteristics: a pair of robins usually has only four young each year, and a pair of mice raises a dozen or more. But their population mechanisms are the same in principle. Both respond to the factors of population dynamics, experience exponential, and logistic growth form, and both are subject to the same rules of environ-

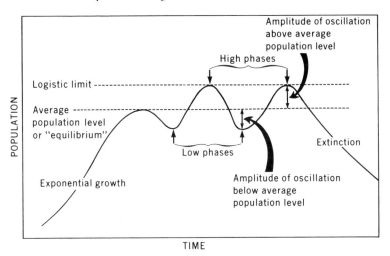

Figure 15. Natural populations vary in number from time to time, but their long-term populations remain rather stable. This phenomenon is called dynamic equilibrium. The amount of a change in number is called amplitude, and a change in time from one peak to another, or from one low to another, is called phase. Average population level is called equilibrium.

mental control. Only the size of increase and decrease varies from season to season, place to place, and time to time. It is the scheme of nature that this should be so.

Every forest is an example of population ecology at work. Forests of the world have evolved into six or eight major types, ranging from the tropical evergreen forests of broadleaf trees to the boreal evergreen forests of needleleaf trees. All of them have developed in response to environmental conditions, especially regional climate (Figure 16). For hundreds of years, foresters have measured individual growth characteristics of trees, as well as the community relationships of forest stands. The "stand table" is one of the commonest features of

A. Even-aged hardwood stand of aspen. Aspen trees are intolerant of shade so the crowns compete for a place in the sun.

B. Even-aged white pine stand with crown-contact characteristic, similar to aspen.

C. All-aged hardwood stand of beech, birch and maple. All levels in the forest are occupied by these shade-tolerant species.

D. All-aged softwood or evergreen stand of spruce and fir with same crown-class characteristics as the beech, birch, maple stand.

Figure 16. Forest trees grow in two general groups of age classes—even aged (or intolerant of shade) and all age (or tolerant of shade). Both evergreen and hardwood species have these characteristics.

interest to foresters, because it is a measure of the value of the wood in a forest. It is also very much like an abridged life table, as well as a tabulation of the age structure of a population of trees. If we sample a large population of trees, we find that all ages are represented —a pyramid of ages. And when we consider that most species of trees have existed for thousands of years and that the lifetime of most trees is less than a hundred years, it becomes apparent that a continuous replacement of one individual tree by another must take place, or the particular species would become extinct.

Some species of trees begin their seedling life in very dense populations. Aspen seedlings, for example, have been found to number about 90,000 per acre. This is about two seedlings per square foot—not really an alarming density for seedlings. It is obvious, however, that as time goes on and the trees grow in size two large trees could not continue to occupy the same square foot of soil; but as growth goes on, mortality takes a toll from each age class. An age pyramid among trees is generated naturally, therefore, just as it is among human populations. Trees, then, like all other living things, are subject to the rules of population dynamics; age structure in a forest is maintained by seed production, analogous to birth rate, and mortality. Obviously, since trees don't "migrate," in the same manner as humans and animals do, the third factor does not apply to forests. Migration is represented only by seed dissemination, sometimes via the wind, other times via water currents, and still other times via animals. Environment determines the success or failure of seed germination, of course, as well as the survival patterns of tree populations.

There are two broad kinds of population structures among forest trees, and they are derived from their evolutionary adaptation. Plant ecologists recognize species that are intolerant of shade and must grow in full sunlight and another class that is shade-tolerant. While intolerant species must grow in stand units all of about the same age, called even-aged stands, shade-tolerant species, such as tropical and all mature forests, have the capability to grow in all age classes. On the other hand, the intolerant species must allocate the available and suitable space to even-aged stands, and therefore each age class must occupy a different stand or site. It follows that, since nothing lives forever, there must be a continuum of age classes on different sites of even-aged stands of intolerant species, whereas for tolerant species, all ages can be perpetuated on a single site (Figure 16). From this contrast, we can see that there is a natural arrangement of trees in space according to their environmental requirements. Trees that are intolerant, such as aspen or pine, are usually only temporary occupants of the site on which they grow and in time are replaced by species whose requirements are conducive to the long-term climatic conditions of the region.

Trees live longer than other kinds of plants, occupy a particular space longer, and therefore give us a better chance to observe their population dynamics; and a stand of living trees, with its reciprocal pattern of survival and mortality, represents all the population dynamics shown in a life table made from a census. Over the long run, this reciprocal relationship maintains age structure in the forest, the same as in all populations, whether they are plant, human, or animal.

Although foresters have studied thoroughly the economic aspects of trees, very little pure population analysis has been applied to forest tree life history. White pine is a tree species that has figured importantly in the economic development of America, and consequently quite a lot is known about it. Although white pine is an intolerant tree and therefore grows as an even-age stand, the mechanism of population dynamics is well manifested in a single stand confined to a few acres of favorable environment. White pine seedlings have been found in a density of more than 90,000 per acre, much like aspen. Over a period of about a century, the seedlings grow into a stand of forest giants, about three feet in diameter and more than a hundred feet tall. By this time, their environmental requirements for space,

light, and moisture will have demanded a natural thin-
ning to about twenty trees per acre.

Mortality over the one-hundred-year period, then,
will have reduced the density from about 90,000 seedlings per acre to about twenty mature trees. Two out of
every 9,000 trees survive to become centenarians,* only
about two-hundredths of one percent of the original
population. But by this time, also, shade-tolerant species
would have filled in some of the gaps in the forest
canopy, and in time the white pine trees would be replaced by some other species. Fortunately, over the
century, such a stand of white pine would have produced billions of seeds, some of which would have
"migrated" to other sites by wind dissemination, thereby
perpetuating the species.

Sea gulls, obviously quite different from pine trees,
were the subjects of studies showing that the same
principles of population dynamics apply to animals as
well as to plants. About 1950, two ornithologists, each
on a different side of the Atlantic Ocean, began a study
of the population dynamics of a common bird, the
herring gull. These two scientists worked independently, neither of them knowing that the other was
engaged in similar work. At about the same time that
Dr. Paludin began "ringing" gulls on some Danish
islands, Dr. Paynter began "banding" gulls on the New
England coast. ("Ringing" and "banding" are the European and American words for marking birds by putting
identifying bracelets on their legs.) Several thousand
nestling gulls were marked and freed by Dr. Paynter
over a period of thirteen years, and more than twelve
hundred marked birds were recovered. Hence, because

---

* Although a hundred years is a ripe old age for white pine, the
species has been known to live 300 years or more.

the time and place of death could be determined and the year of birth was known, their ages and movements could be recorded easily. Having a record of the length of life of so large a number of subjects made it possible to develop life tables that revealed their population mechanism (see Chapter 4). Both dynamic and composite types of life tables were readily constructed, and from these the usual vital statistics were calculated.

Paynter found that the average length of life (mean expectation of further life at hatching) for the herring gull was only about 2.4 years. On the other hand, some birds managed to live for twelve years. This "environmental longevity" of the species can be compared to that of birds kept in captivity for thirty-five years. Since these caged birds were protected from the stresses of a natural environment, it is presumed that thirty-five years represents the physiological longevity of the species; that is, the upper age limit at which the physiology of the body simply wears out. Paludin, working independently on the other side of the Atlantic Ocean, found practically the same population characteristics of herring gulls in Denmark.

The population attributes of the herring gulls that were studied by Paludin and Paynter are particularly interesting because they provide our only measure of how a species of birds avoids extinction. Paynter found from the ways the banded birds were recovered that there are about twelve kinds of "death traps" for herring gulls. Some run into lighthouses, others are caught by seals, some get tangled in fishermen's nets. Paynter also counted deaths by causes and by age classes. Young birds often die because of becoming wet to the skin in storms, one of the highest casualty causes. It occurred so early in life that it tended to reduce the mean expectation of life at hatching time to about 2.4 years. All of these environmental hazards are offset by a reproductive rate of about 2.5 eggs per adult pair. This does not seem very high when we consider all the hazards in the life of a sea gull. Like that of the robin, sea gull populations seem to continue at about the same level, year after year. In spite of a virtual exemption from being shot by hunters, robins and sea gulls appear to have quite stable populations.

This raises the interesting question of how animals avoid extinction. The population mechanism by which this is achieved depends, in the long run, upon a favorable environment. The simple facts of the presence of a species, sea gulls or any other living thing, is proof enough of their ability to avoid extinction. But how does the mechanism work? Paynter found that he could quantify a basic idea of the ecologist Chapman, who conceived the idea of a breeding potential that is offset by environmental resistance, thus maintaining a certain population level. Environmental resistance is the sum total of all environmental hazards, such as those encountered by the herring gulls studied by Paynter and Paludin. Biotic potential is a term Chapman used to sum up what happens when the two forces of breeding potential and environmental resistance act together. The idea is abstract, but it describes something that is quite concrete. It may be expressed as:

Breeding potential — Environmental resistance =
Biotic potential

Using this idea, Paynter found that he could measure the factors and express in numerical terms how sea gulls managed to balance births against the many mortality factors that he discovered.

Breeding potential was measured as the average

number of eggs laid by a female, which he determined was 2.5 each year, and environmental resistance was summed up as a factor that results in an average expected lifetime of 2.4 years. Thus, during an average lifetime, a female gull could produce about six young of both sexes, or roughly three females in 2.4 years. This just about balances out at a little more than the number required to maintain the population against extinction. It might seem absurd to go through this arithmetic when it is obvious, by the very fact that we see these herring gulls year after year, that they are maintaining their populations. But we would have only this very superficial knowledge of wild-animal population dynamics were it not for the detailed work of population biologists such as Paludin and Paynter.

The population dynamics of small mammals, such as mice and ground squirrels, are more complicated than those of most other species because of their reproductive characteristics. Many small mammals have several reproductive cycles each calendar year, whereas most birds succeed in nesting only once. While these mammals are either seed eaters or grazers, and thus have a larger and more dependable food base, they also are more vulnerable to the reductive effect of unfavorable environmental factors. Wet spring seasons, for example, may drown whole litters of young mice that otherwise would mature within the short period of a month to join the procreative adult population. If environmental resistance is not too high, on the other hand, an increment to the breeding segment of the mouse population can occur as many as four times during a single summer breeding season. When this happens, population explosions like "lemming years" and "mouse years" occur, and the fields are literally "crawling" with mice.

Everything that happens to cause such explosions is not precisely known, but in broad outline, the breeding season begins with an "overwintering" population that survived the hazards of winter. Within twenty to thirty days of mating, the old females bring forth three to five young. This is a first generation for a particular breeding season. Survival of this first generation of young is always considerably less than one hundred percent, but even so, within two months the survivors will bring forth a generation of their own young. At about this same time, the remaining old adults from the preceding winter population enter the second breeding cycle. By the end of summer, the old adults usually will have died off but will have left two or even three sets of young, the survivors of which—in the meantime—will have produced as many as two litters.

Although this kind of population explosion sometimes happens, there is always mortality and therefore no population ever attains its physiological population potential. Environmental resistance varies: when it is low, population increases; and when high, population decreases. This alters the number of breeders in a wild state, but it has very little if anything to do with the inherent capacity of a species to reproduce in a wild state.

Toward the end of the breeding season, the compounding effect of high breeding potential and low environmental resistance results in a population explosion. The total population of newborn young and four age classes of adults searches the environment for a safe and suitable place to live. This, in the northern hemisphere, is called the fall shuffle; the young of all animal species seek a home where they can survive winter. They do not all make it. Most of the "displaced persons" succumb to the Malthusian controls of starvation and disease, predation, or accident fatality.

Population pressure causes the dispersal of a population, and dispersal results in redistribution. Among all wild animals, spatial organization is a dynamic seasonal event with changing patterns, depending upon the quality of the environment and the pressures of the population itself.

Familiar birds and mammals that we see every day are subject to all of the principles of population ecology, including the functions of optimum environmental factors, as well as those of population dynamics. None of these functions, or factors, can operate without some interaction with each other. Birds and mammals of field and garden respond to population pressure, create population pressure, organize themselves in space, perform as dispersal agents of disease, consume resources, produce resources, and are involved in the great complexity of co-actions, reactions, and interactions that is called ecology.

VI

Human Population Dynamics

The study of human population dynamics has only recently become an essential part of a liberal education. Heretofore, it has largely been left to the professional demographers, sociologists, and to a certain extent, politicians. Nowadays, every good citizen should know something about population dynamics, population structures, and some of the social and economic consequences of the population mechanism. There is really nothing very mysterious about it.

Demography is the quantitative and qualitative measurement of human population dynamics; it shows the fluctuations created by births, deaths, migrations (or how big the population is at a particular time), and gives information on the individual characteristics—for example, sex and age of each individual—which can be counted. In the aggregate, individuals make up populations, and therefore the sum of these individual qualitative characteristics constitutes the demography, or basic composition, of a population. From the counts of individuals in each sex and age group, secondary characteristics, such as birth and death rate, can be derived. These secondary derivations are only two of the dynamic properties of populations; others, such as the life-table calculations for probability of death, expectation of life, and longevity, are called tertiary characteristics (Figure 17) because they are based on a series of calculations beginning with the simple vital statistics, births and deaths.

In everyday conversation we express the most common features of populations, using three basic characters: numbers, space, and time. Taken all together, these characteristics are known as a census and are implied when we speak of the population of some state or country, such as the United States. The size of the spatial unit usually is not mentioned, but the time is specified as some particular year. Then, of course, the number of people in a particular country is the most interesting thing about a census.

As is well known, the object of a census is to provide the basic description of a population in some particular area that is commonly called a geo-political unit. Such a unit, for example, might be a country or any political division of a country, such as a county, province, or town. Censuses are always made "instantaneously," or rather the goal is to make a count as quickly as possible, because the operation of the factors (listed below) of population dynamics can make significant changes in short periods of time. In large populations a given birth rate can mean a larger numerical gain than the same rate does in a smaller population. The exponential growth curve shows that.

For a closer look at the population mechanism, let's consider the dynamic processes that generate structure

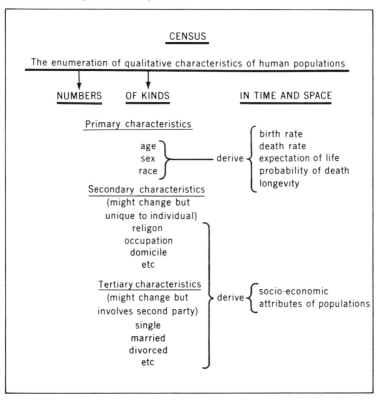

CENSUS

The enumeration of qualitative characteristics of human populations

NUMBERS   OF KINDS   IN TIME AND SPACE

Primary characteristics

age
sex ⎬ ——————— derive ⎨ birth rate
race   death rate
                                 expectation of life
                                 probability of death
                                 longevity

Secondary characteristics
(might change but
unique to individual)
religon
occupation
domicile
etc

Tertiary characteristics
(might change but
involves second party)
single
married
divorced
etc
⎬ derive ⎨ socio-economic
          attributes of populations

Figure 17. Demographic characteristics of populations are derived by statistical analyses of census data. Basically, a census is an enumeration within a definite area which yields a density value, but further analysis of census data yields measurements of other population characteristics, such as sex ratio and age structure.

in all populations—human, plant, or animal. As we saw in Chapter V, there are three main factors that cause changes in populations: births, deaths, and migrations. Gains and losses, from whatever cause, are the basis of the dynamics sometimes called the population mechanism.

The basic *characters* of population should not be confused with the *factors* that cause changes in populations. The relationship between character and the factors that change character is shown in the following tabulation:

Basic Character of Population	Factors That Cause Changes in Character
Numbers	Births
Space	Deaths
Time	Migrations

From this it can be seen that the three basic characteristics enable us to talk about density (how many there are in a given space), distribution (how they are arranged in that space), and changes in density and distribution with respect to time. The three "factors" shown in the tabulation are natural phenomena, which can be counted during a particular time period to describe the way in which a population is changing. Each of these factors can operate at high or at low rates, or, of course, they can remain stable. Theoretically, four possible conditions can result from their occurrence. For example, if births can be high or low, and deaths can be high or low, then the four expected results would be as shown in Table 5.

Although changes in the character of populations are gradual and really cannot be fitted into the nice neat patterns shown in Table 5, these conditions are closely approximated in real populations. Most of the countries of Europe, for example, have experienced these changes over the past 200 years. Any such shift in population structure is called a *demographic transition* and can be of several kinds. The steps in the European transition were practically like the ones shown in Table 5: initial

TABLE 5

Demographic Transitions

Births high + Deaths high = Stable population at a low level

Births high + Deaths low = A growing population with spreading age base

Births low + Deaths high = A declining population; could become extinct

Births low + Deaths low = Stable population, but aging

stages of the demographic transition were associated with high births and deaths, resulting in a small, non-growing, or stable, population. Next came the mechanical revolution and improvements in farming that led to the abundant production of food, which in turn reduced starvation and lowered deaths. Eventually improvements in health and hygiene also reduced deaths. At the time of the mechanical revolution, however, there was not yet a good control over births. The result of high birth rates and lowered death rates was a rapid growth of population. Before this, a more primitive type of stable population was maintained at low density, because a high death rate resulting from starvation, disease, and war offset the high birth rate. This was characteristic of all human populations for perhaps millions of years before the mechanical revolution.

Although birth control in one way or another has been practiced for thousands of years, not until this century did it become effective on a sufficiently large scale to control population increase. More effective birth control was achieved during the late 1800s and early 1900s—after the industrial, or mechanical, revolution had taken hold—and resulted in a stabilizing trend particularly in Denmark, Sweden, Norway, and Finland. Some economists associate this trend with the urbanization of population, which resulted from less need for farm labor. They believe that as a result of the invention of farm machinery, farm labor was in less demand and many people therefore moved to the cities. New life styles in the cities were found to be more expensive, and this higher cost of living is believed to have induced people to have smaller families. Whether this was the cause or not, the transition has in fact occurred in several Scandinavian countries during the past forty or fifty years. City people generally had smaller families than did country people during the past century; but between 1950 and 1965 there was a baby boom in American cities. It appears, however, that, beginning in 1970, a new demographic transition with the lowest birth rate in years is sweeping the country.

In the modern Western phase of the demographic transition, a lower death rate for a period of years, followed by a lower birth rate, created a population that has stabilized at a higher level than that previous to the transition. This is just about the status of Holland, England, Belgium, and a few other European countries today. Scandinavian countries managed to stabilize at much lower levels. Although part of the stabilization was accomplished by emigration to America, voluntary reduction of the birth rate was, in the long run, the most effective stabilizing factor.

For the moment, let us consider certain axioms that can be derived from the interactions of population dynamics. In real-world conditions, a population that is growing rapidly is a youthful population; obviously, in order for a population to grow rapidly, birth rate has to exceed death rate. Therefore, the average age of individuals in such a population becomes "young." Recently settled parts of the world have this demographic attribute—for example, Australia, New Zealand, and Canada are countries with young populations. While some states, such as California, have a younger overall population, others are just the opposite; Maine, for example, Maine, has been "exporting" young people since the Civil War and therefore has a population which, on the average, is both older and aging.

During economic depressions in northwestern Europe, many young people emigrated* to America, leaving the older people behind. Two axiomatic demographic conditions develop from migration: one is that the immigrant population consists largely of young adult males, while the nonmigrant population left behind consists of older people and females. The demographic segment of the American population that settled the western frontier just after the Civil War consisted mainly of young adult males. Quite naturally, young children do not migrate alone; women with children do not usually migrate en masse, except as displaced persons during war; and old persons are inclined to remain in familiar places. This leaves the young adult male segment of the population as the most likely group to move: the miners, trappers, cowboys, and gamblers on the frontiers of the world.

Although the term *growth* in demographic language can mean decrease as well as increase, growth in the usual sense means an increase of population. Naturally, there cannot be an increase without births or immigration. Not even with a reduced death rate can a population grow. Obviously, in order to have an increased birth rate, or even a sustained birth rate, a population must have young adults, particularly women of child-

---

* As noted earlier, emigration from one place results in immigration into a new place.

bearing age. If a population loses the female segment of child-bearing age, growth in the sense of reproduction is absolutely impossible. Therefore, regardless of social taboos or customs, without young women a population ages. It follows from this that the choice of motherhood by young women has much to do with population growth.

Summing up, there are at least three axioms of population dynamics that can be observed in human populations. These are:

1. A growing population consists predominately of young dependents.
2. A declining population consists predominately of old dependents.
3. A stable population is maintained by a balance of births and deaths.

Typical of the first axiom is the Piedmont region of Colorado, in the vicinity of Denver. The fast growth is creating great problems with school financing because of the preponderance of dependent children in proportion to the adult, taxable labor force.

There are only a few countries in the whole world that have come close to stabilizing their populations, and all of them are northwestern European in cultural attributes, as well as in geographic location. Norway, Sweden, and Finland are very close to a balance of births and deaths. Some migrations, both in and out, teeter-totter the balance so that gain, balance, or loss vary from one time to another. Recently, the statistics seem to indicate a very, very slow increase, the slowest of all nations.

A declining population takes place because of a lack of births, or by emigration, both of which result in a residual population of old people. In the case of Ireland during the potato famine, the young adult males moved, leaving the young females and elderly people. Celibacy, delayed marriages, and the absence of marriageable young males led to a declining population that tended more and more toward a preponderance of aged females.

Some demographers and economists are beginning to be apprehensive about the aging of population. Americans now have a mean expectation of life of more than seventy-two years. The tendency to lower retirement age, coupled with increased longevity, means that the old, dependent group will become a larger and larger segment of the population, a potentially great welfare burden on the residual labor force. Japanese industrialists are just now beginning to worry about the effects of reduced birth rate on the availability of a potential labor force. So far, they are concerned only about the loss of a labor force, but aging of the Japanese population would be concurrent with a reduction of the number of people in the working age group. Although these conditions can result in political and economic changes, there need be no concern if lifestyles are changed to fit demographic change.

The process of aging in many advanced societies is also associated with a differential death rate between males and females. In most populations, males die earlier, leaving a preponderance of older females, even though there tends to be more births of males than of females. Males in most primitive tribes face higher environmental hazards from the risk of earning a living. The Eskimo lifestyle, for example, requires that men hunt seals from kayaks in ice-packed sea water. This is a risky business! In East Africa, Masai tribal law demands that young men must kill a lion in order to become a Moroni, a warrior, and while most survive this test of manhood, the idea shows the risks that males take for what sometimes seem to be foolish reasons. Although high male death rates among primitive people seem to be associated with environmental hazards, other environmental factors, such as disease and parasites, are very common causes of infant and juvenile deaths. In any event, deaths from all causes seem to be consistently higher among males of all ages than among females. Partly because of this, polygamy is common among primitive tribes, and therefore high birth *rates* are maintained.

In upper-age classes of industrial societies, a higher rate of heart failure and other degenerative diseases among males appears to be the reason why male mortality exceeds female mortality. Some industrial occupations, too, such as mining and forestry, are high-hazard specifically male occupations.

Thus, higher sex-selective mortality of males results in a higher proportion of females in the older segments of our society.

Fecundity and Fertility

Many people are confused about the terminology used by demographers to denote two important attributes of population—fecundity and fertility. Each species of animal has a different inherent pattern of reproductive capability. While some chickens, for example, have been known to lay an egg a day for many days on end, an elephant can give birth only once in three or four years. This physiological capability to reproduce is called fecundity, while the actual reproduction—giving birth to young—is called fertility. Fecundity is quite inherently, or genetically, fixed and appears to be immutable. Cattle fecundity, for example, has not been changed one whit by hundreds of years of animal husbandry and a century of veterinary science. Cows still can produce only one or two calves per year. The basic reproductive physiology of cows, chickens, goats, horses, and all other domestic animals has not been speeded up in the least, despite the great scientific advances in animal science over a period of many generations.

Human beings also have a basic, and apparently unalterable, physiological capacity to reproduce. The most obvious limitation is the gestation period of nine months. It takes time to make one! Further, it takes time for the human being to attain sexual maturity. Next, there is a period of only about thirty years when a female is fecund. From these simple but fundamental observations, it is obvious that the limiting factor in human population dynamics is first and foremost vested in the human female, even though the decision to reproduce, of course, is altered by social customs and by economic considerations. The fecundity of the human female is the ultimate determinant of human population dynamics and the speed, or rate, at which demographic transitions can be made.

Although it can be argued that various characteristics of the male segment also can be limiting, these limitations are much less likely to have an effect on population dynamics than the physiological limitations of the female segment. Even with higher male mortality in natural situations, the potential possibility that all females will be fertilized is quite adequately assured by the basic reproductive physiology of the species. On the other hand, there probably never has been a complete realization of human breeding potential during the whole history of the human race. Various social pressures, such as religious rites, and economic pressures even in primitive societies operate to curtail the maximum physical capacity to beget children. Thus, for whatever reasons, the actual rate of births stands in contrast to the physiological capacity to reproduce, or fecundity; this actual rate is generally referred to as fertility.

One must recognize that changes in human populations are subtle, slow, and continuous. After all, the most effective factor of change is birth, and the basic birth process must await the maturing of the female, which is at about fifteen years of age among human beings. Nevertheless, the results of change are sometimes observable, as in our crowded schools today. It took the baby boom of the late 1940s and early 1950s to crowd the schoolrooms today. Now we have to pay up! The dynamics of human populations are not really "explosive" as those of some animal populations are. The "ups and downs" of population trends are regulated by basic reproductive physiology as well as by economic and social conditions, although it is not always clear in each "up" or "down" which factors are operating.

Economic Consequences of Population Dynamics

When we speak of national populations, we generally think of total population size. National demographic character is more important than size, however, and it is shaped by birth, death, and migration rates, usually those of the previous generation. Economists and sociologists group any population into three main classes that can affect the population mechanism: the young dependent, the old dependent, and the labor force.

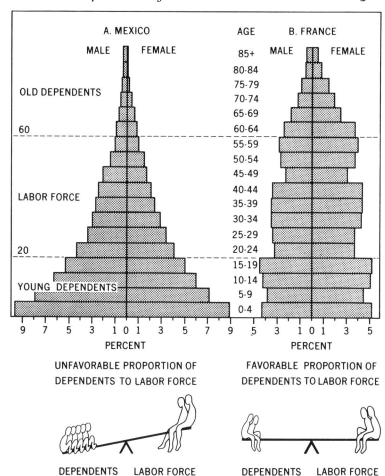

Figure 18. Age pyramids can be divided into three socio-economic groups, as shown in this comparison of the populations of two countries. Young dependents, labor force, and old dependents are found in all populations, but the proportions of these groups vary from country to country. In populations of underdeveloped countries, *A,* the dependents outnumber the workers and consequently the per capita share of the workers' productivity is lower than in *B,* where the dependents and workers are balanced proportionally.

Together, of course, these three groups make up the whole population, and therefore their proportions are reciprocal. For example, if a population consists of half labor force (parents) and half children, there can be no old dependent group. If all three groups are equal, then each is one-third of the total population. All three of these categories can be shown simultaneously by dividing an age pyramid in three parts (Figure 18). This permits plotting the demographic character of national populations so that their economic and social differences can be interpreted and compared.

Age pyramids divided into these three age groups give an idea of the dynamic processes that have shaped populations and the economic conditions that have been generated by their structure. Of course, everything implied in any age pyramid is really hindsight, since it takes time to acquire and analyze the census data needed to make one. Also, the impression of a static demographic condition is implied, even though the factors of population dynamics are always at work, constantly restructuring populations. Births, deaths, and migrations determine the proportions of young dependents, labor force, and old dependents in any population, and the changes in these proportions have profound effects on the human economy and social form.

VII

Ecology of Population Distribution

Distribution of human populations often is thought of very simply as either urban or rural; sometimes, perhaps, it is just "people in foreign countries." Human beings have developed distribution patterns that sometimes appear to be artificially organized, but many of the marks of natural population ecology are still visible on closer examination.

When a census is taken, the principle of the counting system is that each individual has a home base. The entrance to a ground hog den is quite like the door of a house. Around the den there is a grassy area, the home range, in which the ground hog forages. Around a house, if it's in a town, is an area that includes a grocery store, a place of business where the family living is earned, and perhaps a theatre. This pattern in space surrounding the home is the human's *home range,* and the house is his *territory*—quite analogous to a ground hog's den. To count the individual, we identify him with his house, his territory in the ecological sense.

As noted in Chapter I, population distribution depends on environment, and environment has a great deal to do with population density, as well. Density varies according to how favorable the environment is and, to a large extent, governs population pressure. Dispersal from a density center is a result of population pressure; in turn, dispersal reduces density, relieving population pressure.

The organization of living things in space generally is quite closely related to the methods they use to exploit their environment. Beyond that, there is also a seasonal distribution pattern generated by population growth. In spring, robins and rabbits are most numerous and crowd the hedgerows and orchards. By winter they are scarce and scattered. Spatial organization and the population mechanism are inseparable; despite complexities, each must be considered in context with the other.

As a factor of population dynamics, migration causes increases or decreases of population in a particular area, depending upon which way the flow goes. Because it operates in two ways at the same time, it provides a good, simple example of how spatial organization and the population mechanisms are correlated. As we noted earlier, movement of an individual or group out of one area axiomatically means movement *into* another and thus, spatial reorganization. Unlike some human migrations that have resulted in drastic changes in distribution patterns, the migrations of other animals and plants are dependent on optimum environmental conditions. Of course, the ability to live in suboptimum conditions is facilitated by technology, which permits the crossing of natural boundaries and successful re-settlement in unfamiliar environments.

Emigration usually is motivated by push factors out of the country from which the migration takes place, and by pull factors into the country to which the population moves. This more or less is an economic and political viewpoint, although ecological push-and-pull factors operate in the same manner. Unemployment in one locality and the hope of work in another is a common cause of human migration, just as political and religious oppression also have been powerful causes. All of North and South America have been settled by people from the Eastern hemisphere reacting to one or another of these factors. It is noteworthy that migration never took place in the other direction—the American Indian stayed home!

Dependence of Plant Distribution on Basic Environment

Some patterns of population distribution are stringently controlled by environmental factors. Vegetation patterns are easily observed and their environmental causes usually quite obvious. Timberline on high mountains, for example, shows the mutual character of boundaries: a boundary for one kind of distribution automatically serves, also, as a boundary for another. In the case of timberline, the forest exists on one side and tundra on the other.

Plant community boundaries more clearly reflect environmental factors than do the unseen boundaries that seemingly control the movements of animals; plants are rooted in one place, of course, their boundaries determined by environmental factors that affect plant growth. The shoreline of a lake is also a boundary. Such subtle boundaries belie the complex ecological functions that cause them. Optimum environment plays an all-powerful role, and since its principal function in distribution was discussed in Chapter II, we need only to be reminded here that the obvious interrelationships between population and optimum environment are, in the first place, a matter of presence or absence of life in favorable or unfavorable environments. Plants, because they are fixed in space, or rooted in the ground, die if a factor of the environment is critically unfavorable. Therefore, a boundary is automatically formed around the distribution pattern, or range, of that plant species.

The manner in which technology can alter, or modify, life's basic dependence on environment is illustrated by an Antarctic expedition or a moon landing, where human beings temporarily inhabit a grossly unfavorable environment for very short periods of time. Actually, the elements needed for survival in such places must be produced in and transported from favorable environments elsewhere. The principle of this idea applies even to the transportation of plants and animals from their native environments to foreign ones. It is quite probable, for example, that if people did not maintain a favorable environment for them, dairy cows would become extinct in much of their present geographic range. The same is true of many agricultural and ornamental plant species. All of the country now planted in corn would revert to native plants were it not for the continued maintenance of an artificially favorable environment by means of agricultural technology.

Animal Mobility Allows Some Choice

Space utilization by wild animals is different from that by plants. because animal mobility permits a certain amount of choice of environment. Quite often among wild animals, different generations shift from one geographic area to another, seeking favorable environment; for example, the succession of beaver dams along a stream demonstrates the "migration" of successive generations seeking a favorable place to live. Almost always, this kind of migration is initiated by an "eat out," the overconsumption of aspen or willow in the colony's home range. The animals then move to a new favorable location and repeat the exploitation of aspen trees and willow, their preferred foods. They apply their own technology in order to make the maximum use of the area resources, by building dams higher and higher each season to flood a larger and larger area, and by digging canals to float aspen limbs where they can be sunk into the deep water of the pond and therefore protected from winter freeze. Thus, this primitive but effective technology permits a beaver colony to occupy an environment that would be temporarily unfavorable during the winter freeze.

When the food supply is exhausted, however, the animals must migrate—an example of economically motivated migration and of population redistribution resulting from an environmental limit. After a pond is abandoned and there are no animals to keep the dam in repair, a spring freshet frequently breaks the dam and cuts a new stream channel. In time, backwaters form a silt bench and eddies form gravel bars. Within a year or two, a succession of plants invades the site.

Usually, sedges are first; then grasses appear on the silt benches. Alders, willows, and finally aspen sprout again as the site is progressively altered by each succession of plant species. These are each in their turn separate populations that invade the site because—and only because—it is favorable for them. As each kind of plant population matures, the micro-environment of the site is altered to the detriment of that particular plant species by overpopulation and overutilization. The new set of environmental conditions provides a different kind of optimum environment, and therefore the next stage of plant succession occurs: a different species of plant invades the site. In about twenty years, a new stand of aspen usually appears and the site is ready to be reinvaded by beavers. If this happens, and it often does, the new occupants will be about twenty generations removed from the "original" occupants of the site. Thus, beavers are "managers" of their environment, but within the framework of a primitive technology and very definitely within the constraints of a natural ecosystem. Nonetheless, a forester can see the signs of "rotation cutting" by beavers on a twenty-year cycle; coincidentally, this is the principle of sustained yield management in modern forest practice.

A bird's-eye view of a very large area would reveal a definite pattern in the spatial distribution of beaver populations. The underpinnings of the patterns would be seen to be the physical, or inorganic, elements of the ecosystem. First, in order of requisites, is the climate of the area. Water is *the* critical element. The surface geology, or land form, generates a drainage system. Frequently, in most mountainous areas the drainage system is in a dendritic, or treelike branching, pattern. Since beavers are confined to the waterways, their spatial distribution is accordingly influenced by the

drainage pattern afforded by land form and precipitation. Because of physical limits to the system, the population is limited also. But within these limits, a regular
pattern of beaver-colony spacing usually exists on the
waterways.

Position in the Food Chain Modifies
Distribution Pattern

The foregoing discussion of beaver distribution is about
a colonial species of animal that has developed a low
order of technology for exploiting its environment. As
consumers, they are directly dependent upon the producer level of aspen and willow. These animals, like
some primitive farming tribes, are semisedentary. Animals of the secondary consumer level, however, are
carnivorous predators, highly competitive, and typically
poorly organized socially, most of them with no technology beyond the use of shelter. Their highly competitive and independent nature results in a territorial
pattern of spatial organization that is the same in principle but different in detail when compared with beaver
distribution. There is some analogy here with the political territory of primitive warring tribes, who vie with
each other for some vital resource such as grazing land.

The population mechanism and seasonal distribution
patterns of most of the carnivorous mammals are very
similar, even if they do occupy quite different kinds of
environments. Skunks, weasels, mink, and badgers, all
belonging to the same family, are quite different in
appearance but have similar life histories and similarly
designed home ranges, each of which is on a spatial scale
quite in proportion to the size of the animal.

During the dead of winter, when most predators are

fiercely competitive, populations of these mammals are lowest and most stable in spatial organization. By this time, also, many of the young have succumbed to various mortality factors. This is the "pinch" period, the time of keenest competition with others of the species and with the elements as well. Foxes, coyotes, minks, and marten are most likely, at this time of year, to be mostly adult or subadult populations, appearing to claim, or to attempt to defend a home range.

The habits of red foxes are rather well known. A given area of habitable range is divided by a resident fox population into home ranges that are usually about three miles in diameter, as any "hound-dog man" knows! The home range is the natural spatial unit on which the very ancient idea of "hounding" is based. Just a brief description of the happenings in a fox hunt can illustrate the principle of home range: the hunter releases his old experienced strike dog (one with a keen nose), which, upon striking fresh scent, "opens up," "gives tongue," or "bugles" (according to the various collo-quialisms in different parts of the world). At this time the hunter "casts the pack," and these dogs join the strike dog, which is usually slower and surer, giving chase. The hunter, of course, is not able to keep up with the hounds, but knowing the terrain and the habits of foxes, he hastens to a "crossing" and waits for the hounds to put the fox by him. The only reason that the fox comes by him, in all probability, is that it runs familiar ground, the home range. This principle is used in hounding all game, from rabbits to moose, with hounds of various sizes and speeds, from bench-legged beagles to Norwegian elkhounds. There is a very close correlation between the size of the game and the size of the home range. It follows, then, that there is a density limit for each separate species, described earlier as the "carrying capacity." There is also a particular distribution pattern determined by the favorite environment of the different kinds of animals.

Realistically, the distribution pattern of each different species is influenced in the first place by favorable country, which embodies the idea of optimum environment. Terrain also influences the shape of home ranges almost strictly in accordance with shape, size, and juxtaposition of favorable habitat. Because of this, home ranges are only roughly circular, with a radius that is quite proportional to the size and mobility of each species.

Shown in Figure 19 is the relative size of the home ranges of three species of carnivorous mammals, all of which compete—in part—for the same foods, but each of which has some structural advantage to exploit some particular food source to the exclusion of the other. The figure idealizes the home ranges as circles, and each circle is proportional in radius to the home range of the species. It follows very simply that the smaller the home range, the denser the population within a given area.

Consider now the matter of relative density: density is proportional to the carrying capacity of the optimum habitat, which is relative to the body size and food requirements of each species of animal. Here we are reminded of the pyramid of numbers: more small animals than large. Figure 19 can be used to represent the distribution of three members of the skunk family in a wintertime distribution of adults. Each population, however, consists of male and female of the species, and it happens that there is a definite sex-specific difference in the size of home range among all three of these mammals. Competition for food even forces the sexes to segregate during the winter. Presumably, males require more food and therefore a larger foraging area, or

10 MILES

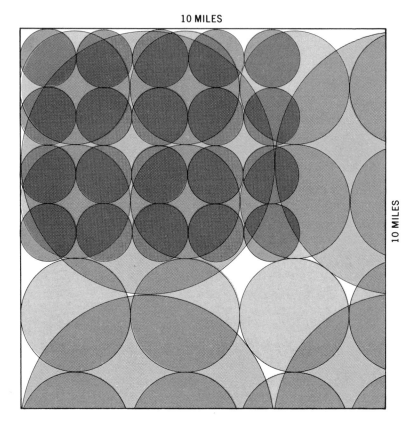

10 MILES

Figure 19. Density and distribution of animal populations depend on size of the animal, food requirements, and size of its favorable habitat. The wolverine, marten, and ermine (or weasel), all of the skunk family, have similar habits but each requires a different home range space because of a difference in size and food requirements. A 100-square-mile area, 10 miles on a side, could be the home of one wolverine, 35 or 40 martens, and more than 500 ermines. Many other animals could use the same space at the same time, each using different resources of the area or competing for the same products of their environment. (The large circle represents home range of a wolverine, medium circle of a marten, and small circle of an ermine.)

home range, than that required by females. This does not alter, however, the basic arrangement of a pattern of circular home ranges distributed over a field of optimum environment.

Perpetuation of the species, of course, depends upon a breakdown of this pattern of solitary ranging, and the two sexes do break down this distribution pattern temporarily in the spring breeding season. These particular species do not share the work of raising young, so the segregated pattern for the most part is resumed after the breeding season. Foxes, coyotes, and other members of the dog family demonstrate this same solitary pattern during most of the winter, but during the breeding season the males and females generally pair and share the burden of providing food for the young.

Among fisher, marten, and ermine in the spring, young are born roughly in a ratio of about four per female. This sudden exponential increase crowds the range, but fortunately the whole ecosystem is geared to it. The "green up" of grass, the "explosion" of mouse populations, and the return of migrant birds that multiply in accordance with the exponential curve are interdependent parts of the whole ecosystem. They each feed on the system and, in turn, are themselves fed upon.

Land space is now overcrowded as each kind of population attempts to express its inherent power of exponential growth. But the forces implicit in the logistic curve work on all populations—plants, humans, and animals alike. During the summer, the young of all kinds of birds, mammals, fish, and other animal life feel the pressure of population increase, the "squeeze" of environmental limitation. A "shuffle" of population is unavoidable, with the young and naive being most

vulnerable. There are more of them than there are
adults, they are less wise to the ways of their world,
and they are therefore more vulnerable to the factors
of mortality. Hence, the law of chance picks on them.

By the time summer ends, the young of foxes, coyotes,
and their cousins of the secondary consumer level, segregate themselves from each other and from their
mothers and range for themselves, each seeking a niche.
While some are successful, many fall victim to the
hazards of life. By early winter, a pattern of distribution
is redeveloped that fits the maximum carrying capacity
of the environment; that is the "saturation point," which
varies from season to season as well as from year to
year under the control of favorable or unfavorable environmental factors. In such a fashion, an annual variation of distribution pattern occurs, a pattern of density
that is dependent on the most favorable environment.
Thus, the distribution of wild things is not static. It
varies from season to season with the population mechanism and from year to year with an amplitude that is
controlled by either good or bad times, as afforded by
the physical factors of the environment (Figure 20).

There is a certain quality of the "territorial impera-
tive" demonstrated by wild animals, although biologists
prefer to reserve the word *territory* for use in describ-
ing an area related to breeding activity, and *home range*
for feeding. Some kinds of animals distribute themselves
as solitary, highly competitive individuals; others, such
as the beaver, are gregarious or colonial.

Among the colonial animals there is a degree of
group-area sovereignty, and the range rights of the
individual are either nonexistent or obscure. A herd of
African game is really several herds made up of different species; all apparently utilize the same space,
but in reality each uses only those micro-sites that

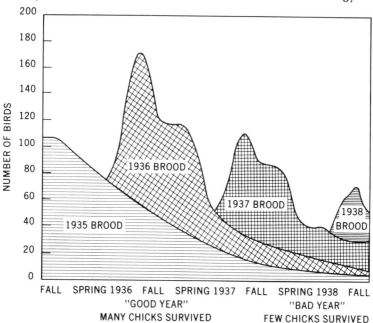

Figure 20. Populations of valley quail change with good and bad seasons. Favorable or unfavorable environmental conditions account for the amplitude of populations as well as the changes in sex ratio and age structure.

proffer the specific kind of forage that it prefers. Within each species herd there is some kind of family unit herd. Among elephants, for example, the matriarchal unit usually consists of four individuals: the old mother and three young calves, which are spaced in age about four years apart because of the long gestation and long nursing period. As new young are born into the population, roughly at four-year intervals for each individual female, the oldest of the young animals becomes mature, at twelve to fifteen years of age, and breaks away on its own. At times, an "old auntie" appears to attach herself to such a unit; this is part of the gregarious pattern of herding. In such a way, the usual herd of elephants consists of multiples of four, or nearly so, and the most common herd size where elephants are abundant is sixteen. Although these herds have favorite haunts, organized as spatial home ranges, their trails connect with home ranges of other herds and from time to time they assemble in huge herds of a thousand or more for the purpose of breeding.

Spatial organization among some birds takes on a three-dimensional aspect. In addition to the usual idea of space on a two-dimensional plane, there are species of birds that nest at specific height levels in trees. Usually, small warblers nest within ten or fifteen feet of the ground. A robin might nest in the same tree with a warbler, but at a higher level. Most birds demonstrate territoriality around the nest quite vigorously and will fend off birds much larger than themselves. On the other hand, these same birds, of different species, would have no conflict of interest in the spatial home range because their environmental requirements differ, and the particular space that several species might be using at the same time affords the specific requirements for each. That is patent in their presence! It is also evident that the same space can be the optimum environment for two quite different kinds of animals at the same time.

Much more could be said of the life histories of many kinds of animals that are so wonderfully different in appearance, but the basic picture of spatial distribution among all wild animals sticks quite closely to the idea of territory. To quite a considerable degree, the phenomenon of territoriality is the smallest spatial unit in patterns of animal distribution. Next in order is home range. A third higher ecological space division is optimum habitat, the arrangement of all the resources—

food, water, and shelter—that are required by any organism unable to manufacture its own food. The fourth level of spatial organization is an extension of the third—ecological range. This is the collage of optimum environments within the geographic range, and each might not necessarily be contiguous with others. Last in the hierarchy is geographic range, the maximum geographic extension in space where a particular species may be found. It is controlled by gross geographic phenomena such as the climatic pattern where it is feasible for a particular plant or animal to live.

Some of the spatial requirements of animals that we have discussed are likewise visible in human ecology. The home equates with territory and the idea of private property rights. The area where a person lives, works, and plays is essentially a home range. The land space occupied by a group of people with a similar language and similar customs is much like the geographic or ecological range of some species of animals and may be compared with the idea of national sovereignty. Although national boundaries more or less have been stabilized and formalized by cultural development, among many groups, even today, the biological attributes of home range and territory are still readily perceived.

VIII

Human Settlement Patterns

People are scattered or crowded together more or less according to an economic order of development. Even the cities of India, which are crowded with poor, have their cores of commercial and industrial activity. Here again, as with animals, there is a scale of distribution that is related to environmental carrying capacity, but the difference between wilderness and megalopolis is the product of the cultural enhancement of environment. People in rough and austere environments, such as the mountains of Peru, or the deserts of Arabia, or the arctic barrens of Baffin Island, are scattered in very small clusters over very large areas. Population densities of such places are low because their natural carrying capacity has not been increased by the application of a high order of technology. Inhabitants of areas such as these are generally nomadic and must wander over large areas in order to glean a livelihood from the low productivity of their native environment. That is why many American Indian tribes were nomadic hunters.

Peoples such as the Lapps, Eskimos, Tungus, and Samoyede have developed particular cultural adaptations to a particular set of environmental factors but have not yet developed a political sovereignty. Their population characteristics, especially high death rates, act to suppress population pressure, and therefore the need for any form of "national" expansion never gets very far. Usually, very high death rates occur in these inimical environments, and here again we see some impact of basic environmental factors on the population mechanism. Nevertheless, definite settlement patterns develop, even in sparsely populated regions.

Primitive settlement patterns and ancient ones must be recognized as quite different. The ancient city states of Greece were population aggregations of considerable size. Concentrations of population require a constant flow of food from a hinterland, and the Greek countryside was dotted with patrician estates which supplied this food. So the basic pattern of ancient Greek settlements was quite like the patterns of towns and roads found in Europe and America today.

Patterns of population distribution always have been influenced by both environmental and cultural factors. Birdsell's studies of Australian aborigines is a very good example of environmental influence on primitive population distribution (Figure 21). Two main features are illustrated here: the density of a human population is limited by a natural resource, water; and the distribution pattern that results is typical of people who have not developed a technology to modify their environmental resources. A rugged desert environment and a low-order technology have combined for thousands of

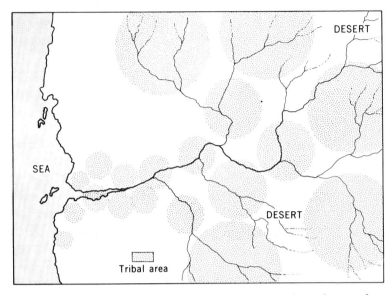

Figure 21. The spatial distribution of Australian aboriginal tribes is determined by water, the environmental factor in this desert region. The closer a tribe is to the river mouth, the larger the volume of available water and therefore the smaller the tribal area need be. Consequently, populations are more dense near the mouth of the river than in the interior deserts.

years to prevent the population from growing to the point where tribal conflict over territory is of any consequence. The pattern of tribal territoriality has been arranged in space in such a manner as to utilize all of the available water supply without any cultural modification. Tribes with larger water supplies require less space, and tribes with less water require more space.

Birdsell has recognized two classes of population aggregation among Australian aborigines: clans have a close-order family relationship and move about independently within the tribe's area, and all of the clans of one resource area belong to a tribe. The various tribes are associated with water supply areas, and because of this, there is a low-order tribal territorial pattern that fits the biological description of a home range quite well. As is well known, such traits have never led to the organization of a political or economic framework other than the rudimentary tribal economy of gathering a livelihood from the land.

There are many examples of low cultural orders and consequent sparse population patterns, even today. They can be found in just about all of the biomes or ecosystems, with greater or lesser degrees of cultural development. In the tundra biome, which has a circumpolar distribution, there are Eskimos in the American tundra and there are Samoyede, Tungus, Chukchi, and a few other groups in the Asian tundra. Deserts have wandering tribesmen such as the Bushmen of the Kalahari and the Tuareg of the Sahara.

Judging by the American Indian tribes that are scattered from Tierra del Fuego to the arctic coast of Alaska, few of them ever developed their area resources beyond the level of hunting and gathering; consequently they did not expand their spatial distribution beyond their natural environmental boundaries. Highly developed artistic and cultural patterns among such groups as the ancient Inca, Aztec, and Mayan populations—and even among modern Hopi, Zuni, and others—were not strongly exported to other areas and remained more or less in indigenous core areas. There seems to have been some trading, but quite definite cultural core areas seem to have remained spatially static for very long periods of time. This suggests that there was no population growth and, therefore, no territorial expansion.

Even though the artistic and subsistence technologies

were quite skillfully developed, the human populations of the Americas had not succeeded in achieving any methods of death control. As a result, population pressure was minimized. The extent to which tribal warfare controlled population is not known and probably never will be accurately assessed. Mortality was a factor, there is no doubt about that, but the effect of war on population control appears to have been less than that of starvation, disease, and natural hazards.

The development of spatial distribution patterns of human populations is more complex than that of animals because of the various levels of technical and cultural improvement upon the natural environment. Laplanders have a rather unusual pattern of spatial organization; it results from their dependence upon the reindeer herds, which migrate to avoid unfavorable seasonal conditions. In summer the reindeer move to the seacoast, where it is cooler than the interior tundra. In winter they live on the tundra, which provides dwarf birch, willow, and the lichens called "reindeer moss" that is their mainstay diet. Accordingly, the Lapps must move with their herds. Over the years the Lapps have decided that it is best for them to regard the deer as private property; but because of the migratory and semi-wild nature of the deer, the grazing lands are recognized as communal property. The Lapps are so busy constantly packing up and moving that they scarcely have time to get attached to land and, therefore, have no need for the idea of landed property rights. Nevertheless, each and every reindeer is marked with a specific owner's brand and regarded as private property.

In this connection, the origin of the word "cattle" is interesting, having been derived from an old French word, *chattel*, for movable private property. The Laplander idea of private property rights in livestock and not in the land that supports it is also common among tribes of herdsmen in Africa. Ankole, Acholi, Masai, Karamojan, and others recognize the right of private ownership of cattle, but the lands that support their cattle are communally held by the tribe at large. In contrast, it is significant that neighboring farming tribes such as Bunyoro, Kikuyu, Buganda, and Toro recognize ownership patterns on the land.

To a certain extent, there seems to be some kind of land ownership classification among primitive people based upon intensity of use. Herdsmen using land extensively do not develop a personal land ownership right, but farmers using land intensively do. Both groups, however, recognize tribal regions and have gone to war over them periodically for thousands of years. Thus, we probably have in these cases the most primitive type of land sovereignty.

During the height of the Greek and Roman empires the framework of our present settlement patterns seems to have been in use. Cities, towns, and villages were connected with roads. Between towns and villages there was open countryside and well-developed farmlands. About the same time that the Roman Empire flourished, far to the north the Vikings of Scandinavia had a land-ownership pattern that also has left some mark on our own. The Vikings of this period held council meetings called *tings*. Today, the Norwegian national parliament is called *The Storting*, meaning large meeting, or assembly. In order to have a voice in the ancient *Storting*, a Norseman was required to be a landowner. Ownership of the land implies some form of settlement, and of course this was necessary in order to make any reasonable living from the land. In addition to this, it was natural that the landowner had to be a free man, as a slave could not own property. The free man also had to reserve the right to bear arms in order to protect that property. We can discern some of the basic principles of our own constitution in these ancient customs, in particular, private property rights.

Norwegians developed a valley settlement pattern, which, if you have seen the rugged fjord country, is about the most sensible thing they could have done, considering the rugged mountain and valley terrain of the country. Even today, Norwegian girls wear costumes that are distinctive of their home valley. The old custom of having a summer farm in the alps and a winter farm in the valley can still be seen as a basic pattern of seasonal population movement called *transhumance*. The larger valleys are organized geopolitically into provinces, which are the major political divisions of the national estate.

Most national groups have developed a hierarchy of political organization that closely resembles patterns of population distribution. This is visible in such modern English words as town and hamlet. The word town is derived from an old Nordic word, *tuin*, for a garden hedgerow, which is a boundary of a person's property. The word hamlet, meaning a very small village, comes from the word for home, which is variously *heim, hame*, and *hem* in several dialects in northwestern Europe.

Patterns of human population on the landscape in some parts of Europe still reflect the medieval period of about a thousand years ago when great manorial estates dotted the countryside. Many of the old manor houses still stand as small castles today, and where serfs once lived in cottages on the manor, free men now live but still work the farms as tenants or shareholders. In Christian times the church created another dominant landscape feature that can still be seen in France and Switzerland. A cluster of small cottages and specialized craft shops, such as those of a blacksmith, shoemaker, and weaver, surrounded the church.

It was only five hundred years ago that many castle fortresses flourished throughout Europe from Spain to Scotland. These bastions were not self-sufficient; they could be sieged, as many were over and over again. Outside the castle walls there had to be a productive countryside, controlled by the lord of the manor, but with people living on the land and working it. This pattern of population distribution still exists, although the political and economic status of the people have changed several times over.

At one time in most of Europe, there was a registry of nobility that included a landed title. The noblemen were called the "landed gentry." Although this did not completely control the pattern of human population distribution, there were some powerful influences on settlement patterns that are still visible on the land today. In most of the countries there was an equivalent hierarchy of noblemen, but the words used for them naturally were different in each different language. In the English system, for example, a *barony* was a considerable tract of land owned by a baron. A count owned somewhat more land, the term being the base of our word *county*. Dukes presided over a duchy, which in England is still a major political land division, such as Cornwall or Northumberland.

Our picture of the development of population distribution patterns is not very clear much before the medieval period of about 700 to 1200 A.D., or roughly a thousand years ago, when it appears that agricultural technology began to be effective in improving food production. Prior to that, tribal or clan patterns of habitation seem to have been world-wide in scope. America, for example, both North and South, appears to have been populated from one end to the other, even though sparsely. When Columbus landed on the island of Santo Domingo in 1492, he found native human beings who spoke a particular language, Carib, and had customs befitting the environment. Only five years later, John Cabot landed in Newfoundland, two thousand miles away from the point of Columbus's landing, and found native people that spoke a different language, Micmac, and had customs adapted to their particular environment. They were quite different from the Caribs, and it was quite obvious that they had not traveled two thousand miles from Santo Domingo in the short space of five years! Only ten or fifteen years later, Spanish explorers crossed to the Pacific and spread both north and south along the Pacific coast. They found the whole country populated with groups of people that spoke different languages and used different natural resources in different ways. In other words, the whole continent was inhabited, and each distinct ecological unit was maintained as tribal territory as long as the population pressure was not so great as to cause territorial expansion.

Occupancy of the land in Europe during ancient times appears to have been quite similar to pre-Columbian patterns in America. We read of Celts and Picts in the British Isles, which seem to have been peopled in a more or less tribal spatial pattern over the landscape. Only recently woodhenges have been discovered, whereas we have known stonehenges for some time. Along the River Avon, woodhenges found spaced about thirty miles apart appear to have been population centers. Although they certainly indicate human habitation and offer some clue to population level and distribution, not much has been worked out about them.

Different tribal names signify a distinction in geographic distribution. Ostrogoths, Visigoths, and Vandals, some of the better known ancient tribes of Europe, all occupied distinct geographic areas, even though at times during their histories they moved about to conquer the peoples of other areas. Although we believe that populations were sparse, there is ample evidence of widespread habitation during ancient times. It seems that all of the land space of the world has been occupied at one time or another.

Nearer to our times, the journals of Marco Polo give a very interesting perspective on population distribution. Traveling extensively in Asia during the period of about 1260 to 1280, the latter part of the Medieval period, he recorded notes about great cities surrounded by smaller towns, and towns surrounded by "posts." He described a highly productive countryside with settlements spaced about a day's horseback ride apart. In fact, he records a kind of pony express that the Great Khan maintained, and post stations where a change of horses could be made. In reading his accounts, one gets the impression of a traveler going horseback from settlement to settlement, never more than five or six miles from some population center. His accounts distinguished these peoples from himself according to appearance and custom. But the distribution of the people on the land was not much different from the node and network pattern of town and road in his native Italy. This land produced wheat, among many other crops, to support the extensive population. The Mongols were great meat eaters and had huge herds of goats, sheep, and horses, but they also made a dish out of wheat that Marco Polo introduced into Italy as spaghetti, the Italian "national dish"!

Any atlas shows a considerable communication network in all parts of the world. Even in the deserts of

Arabia and the Sahara there is a node and network pattern of settlements and roads or trails, albeit a very sparse pattern. For centuries, Arab traders with camel caravans have crossed these deserts to exchange goods at points great distances apart. Timbuktu, for instance, was on one of the trans-Sahara caravan routes. And among very ancient trade routes, there was the "silk road" across Persia, now Iran, which connected the Orient with the Mediterranean.

The world over, we find a pattern of human population that reflects two principles of population distribution in space: the first, *central place* theory, holds that the spatial distribution of human settlement consists of population clusters connected by a network and arranged around a node, called a central place, which is larger than surrounding settlements. The German geographer Christaller observed that this arrangement generally exists as a hexagonal field with a fixed number of minor localities regularly spaced around the central place (Figure 22). This can be tested on a map of roads around any home town, but scale must be considered; the network in the model is airline, not road, distance.

The second principle of spatial distribution, called *rank size*, deals with the comparative sizes of nodal populations—the central place, which, naturally, is largest, and the surrounding clusters of population, such as towns and villages. There is a regular hierarchy of city (or place) size. These are reflected in such English words as megalopolis, city, town, village, and hamlet. Most languages have equivalent terms, which attests to the world-wide occurrence of this pattern of towns and villages around a big city. The largest city of a region is called the primate city and is designated as P_1. The rank of each surrounding town or village is

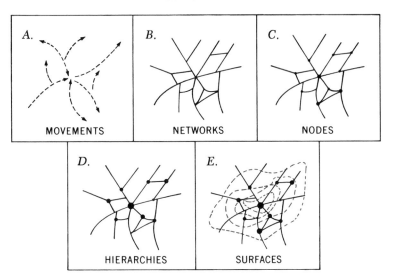

Figure 22. The node and network surface pattern of population distribution is called Central Place Theory. This theory holds that the surface is a hexagonal field with about six points, or nodes, of population surrounding a larger population node called a central place. *A.* People naturally need to move about in their environment. *B.* Eventually, a network of footpaths is developed as people travel back and forth in their habitat. *C.* Settlement nodes, such as farms or villages, develop. *D.* Some settlements grow more rapidly than others and develop a rank-size hierarchy. *E.* The process naturally takes place on a surface—the land.

determined by comparing its population to a scale derived by dividing the population of the primate city by 2, 3, 4, and so on. In other words, a second-rank city is about one-half the size of the primate city of a particular region, and the third-rank towns are about one-third as big as P_1. It is surprising how well real population data—such as the 1970 decennial census data—fit this scheme.

Once a central-place system is established over a given geographic field, the pattern does not usually change. Experience has shown that each separate place tends to grow independently. Therefore, a regional density may change but the basic central place pattern remains. Furthermore, the space between nodes usually fills in with new nodes, or settlements, such as sub-divisions. London once had outlying towns such as Kensington and Charing Cross that have now inter-grown with the capital city.

Each settlement unit in the hierarchy of rank size has a certain territorial sovereignty. This is expressed in such terms as incorporated or unincorporated village. A social and financial responsibility is embodied in this geopolitical sense. Thus, sovereignty can mean responsibility as well as independence, as it is more often taken to signify.

Through some thousands of years, for various demographic reasons such as reduced death rate, and for various technological reasons such as increased food production, populations have grown, and they have grown according to the laboratory experimental model of the exponential growth curve. With the natural establishment of a central-place system and exponential growth of each place, the land space gradually becomes saturated, resulting in crowding and environmental stress.

These processes have created controversies over territory throughout the history of mankind. Wars are still being fought over this basic biological requirement: the territorial imperative. Out of it develops a certain space-oriented protectionism much like that shown by a bird when it defends its home range from another bird. Human beings find it difficult to evade this basic biological compulsion; even though a landless society is developing from the process of urbanization, people stacked together in apartment houses still require the same privacy that compels a bird to defend its nesting territory. Collective human populations with common cultural characteristics, such as language, express this territoriality as *nationalism.*

IX

Settlement Patterns and the Diffusion of Ideas and Disease

An idea cannot spread unless there is a population distributed over the countryside to spread it. The spread of religion, like that of other ideas, depended entirely on the distribution of a human population and the dispersal of religious ideas by that population. Without clusters, or nodes, of populations scattered on the landscape, there would be no way for an idea to be dispersed. The node and network pattern of population distribution is the medium of dispersal for an idea, as well as for the spread of disease. The diffusion of a new idea is dependent on the density of population: the fewer the people there are scattered through a region, the slower an idea spreads. Conversely, rumors spread more quickly where populations are dense. Disease does too, as every schoolchild's parents know.

When we recognize that populations are nodes composed of individuals, and that these are connected by a network of roads, we can trace the pathway of disease diffusion, which seems to begin with the individual, run through the family and into the community, then along the network—the pathways of communication over land, sea, and air—to other nodes, or communities.

An outbreak of disease invariably begins with one or a few individuals in some settlement, usually in the larger, more densely populated cities. There is a certain mystery about the process. Where does the disease come from? Why does it begin with a particular child in a particular schoolroom? Of course, there are many different kinds of diseases and some of them appear not to be related to the diffusion process by which others spread. But of those that do have these characteristics, it seems that there must be some reservoir where disease organisms incubate, hide out, or otherwise lie dormant until conditions are right for the disease—and wrong for the person who gets it. At least one thing is certain: diseases are subject to all the rules of population ecology.

At this point we must once more break into the ecological circle, assume a place to begin, and thread through the chain almost as if it were a food chain. In fact, disease organisms are part of a food chain. The fact that one or two individuals seem to be afflicted with a disease before it spreads more widely through a population leads to the suspicion that the human body provides the environment where the disease organism finds optimum conditions. That individual becomes the host for the disease and, in turn, passes it on to another.

Disease organisms—whether they be bacteria, viruses, or protozoa—survive in suboptimum reservoirs where they sometimes remain for long periods of time until an environment becomes optimum for them. Many diseases have life cycles that are made up of two or more diverse forms, and therefore two or more optimum environments are necessary for their survival. For example, malaria is caused by a protozoan, or one-celled animal, that spends part of its life in a particular mosquito and part in man (as well as in other organisms); both mosquito and man function as reservoirs for each particular stage of the life cycle. In turn, the specific mosquito carrier, the *Anopheles* mosquito itself, requires an optimum environment. Fortunately for people in Northern countries, that optimum environment is in the tropics and subtropics. For the purpose of recognizing how two or more kinds of populations— in this case, the malaria protozoan and man—are involved in a gigantic ecological system, each with its own population mechanism related to density, distribution, and dispersal, we shall have to examine the function of their population mechanics in an oversimplified way.

The first simplification is to assume that one person is an optimum environment for the protozoan. Incidentally, the word malaria, is formed from words meaning bad air, because it was once believed that the disease came from the bad air of tropical swamps. Indirectly, it does: swamps provide the optimum environment for the mosquito that carries the disease! The second simplification is to disregard the population ecology of the carrier of the disease, the female mosquito, and assume that one of them has transferred the malaria protozoa to some human being. When we think back through the foregoing discussion of population ecology, the thought of the food web or food chain comes up. If some small fish such as a gambusia had been hungry at the right place and at the right time, this particular mosquito would not have lived to transmit malaria. And if this particular mosquito had not been hungry for human blood, the human being would not have become the environment which the malaria protozoan found optimum.

Once the malaria protozoan enters the body, it begins to feed on the red blood cells. Finding a good food supply, the organism reproduces exponentially, its growth consistent with that indicated in a logistic curve, and a fever results. There are several forms of the disease, and they each produce fevers on different cycles. One is on a forty-eight-hour cycle and one is on a seventy-two-hour cycle (Figure 23). Why the cycles? Population mechanisms at work!

The human body produces white blood cells, or

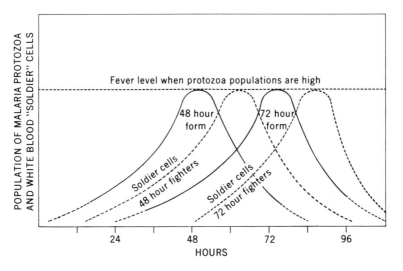

Figure 23. Fever from malaria is a cyclic population phenomenon very much like a prey-predator cycle (Figure 6). Two forms of malaria have two different "phase" periods: one on a 48-hour phase and the other on a 72-hour phase. Nevertheless, the basic principle of the population mechanism is demonstrated by both forms of malaria.

corpuscles, to fight the disease organisms. Again, in order to explain it, we have to break into this cyclical system beginning with a fever. While the mechanism of fever is not well understood, we know that in malaria the fever is associated with a high population of protozoa. Then, the population of white blood cells, increased by the body, attack and subdue the malarial protozoa. Fever then subsides; the human body produces new red blood cells and recovers its health at least to some degree. When the white cells are no longer needed, they are absorbed and their population thus decreases. Shortly, in forty-eight or seventy-two hours, depending on the type of malaria organism, the protozoa population builds up again and the fever returns on a rather consistent time cycle.

A typical population growth curve can be plotted, showing the two warring populations in the battlefield of the human body. In this diagram (Figure 23) we see the amplitude of the protozoan population on a time cycle that is the phase of the fever cycle. The white blood cell cycle is, at least theoretically, on the same amplitude but on a different time phase—it lags. In the time sequence, the white blood cell population does not build up until after the protozoa population builds up and decreases after the protozoa have been subdued.

Realistically, the process involves millions of people, probably billions of mosquitos, and an uncountable population of protozoa. But even one *Anopheles* mosquito is likely to "bite" more than one person and in so doing begin the diffusion of the disease. The more people, the more rapidly the disease can be spread! The higher the density of people the higher the incidence of the disease is likely to be. On the other hand, the spread of malaria, like that of most diseases, is also related to the distribution of all of its host populations. Human population is not limiting because all human beings are vulnerable and have a wider geographic distribution than either the mosquito or the protozoan. But, although humans are carriers, one person cannot transmit the disease to another, directly.

Malaria is an endemic disease; that is, it does not spread geographically on its own but rather from mosquito to person to mosquito. If an affected person moves to a locality where there are no mosquitos to transfer the disease, he cannot function as a vector. But if he remains within the ecological range of the carrier mosquito, he functions as a reservoir and secondary carrier, which facilitates the survival of the disease, as well as its transfer to other people.

The words "endemic" and "demography" come from the same word root, *demo,* which is Greek for people. Endemic means belonging to people and has been transliterated to mean within a given locality. Pandemic means widespread, or "all people," while epidemic refers to a disease that flashes through a population rapidly and is not permanent in any particular locality. These terms relate to the distribution and dispersal of disease and also indicate the role of population density in the spread of disease. Centers which are dense generally function as reservoirs from which diseases such as malaria and cholera can get started on a diffusion pattern that can spread around the world.

In ancient history, we find that the same principles of disease diffusion have always existed. The story of Saint Luke, the physician, includes some accounts of the diffusion of disease around the Mediterranean Sea. Ships sailing from the Egyptian port of Alexandria

carried passengers who were infected with plague. The
accounts tell of some people who died at sea from the
disease picked up in port. Once aboard, they were a
dense population node, with no escape from the close
contact with the passengers who were carriers of disease. Therefore, whole shiploads of people sometimes
perished.

There are accounts of the rate of spread of disease
in ancient times, the direction and speed of diffusion
being exactly traced by the course of a ship. Ships
carrying infected passengers floated around and were
required to fly a yellow flag: they were not allowed to
enter port until all on board had recovered from the
disease. Some ships must have floated around flying
the yellow flag of quarantine even after all passengers
and crew aboard were dead of disease or starvation.
Even so, some of the survivors of plagued ships carried
disease into port, perpetuating diseases through generations, or we would not have the same diseases among us
today.

Several times during the past five or ten years there
have been waves of influenza spread from Asia and
Europe to America. It is easy to imagine an airplane or
a ship bringing one infected person across thousands
of miles of air and sea, like a thin thread connecting
two great population masses. The node and network
pattern on a surface is the disease diffusion system—
not the cause, but the diffusion system. In the case of
human diseases, people, of course, function as an environment for a disease and as a carrier, or diffusion
agent, at one and the same time. The map (Figure 24)
shows the rate of spread of some of the recent flu
epidemics.

DIFFUSION OF INFLUENZA ALONG SHIPPING ROUTES

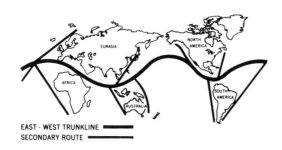

EAST - WEST TRUNKLINE ▬▬▬
SECONDARY ROUTE ▬▬▬

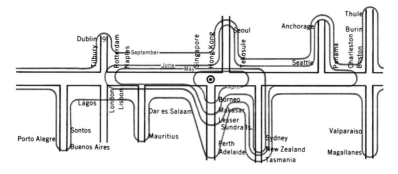

⊙ FIRST INFECTION OUTSIDE CHINA IN APRIL

Figure 24. The diffusion of disease depends upon a network of roads or other means of transportation and communication that connect the nodes of population, and on the movement of people over that network. Ideas, religion, rumors, and disease all spread in the same fashion, known as the diffusion process.

The Geography of Disease Distribution

Fortified with some idea of the population mechanism that is involved in the diffusion of disease, we can now put the horse before the cart and examine the role of the physical environment in the life cycles of all diseases. By no means, however, can we do this in a simplistic fashion, ignoring the population mechanism and the arrangement of populations in a node and network pattern that facilitates diffusion. Neither can we ignore the function of populations as reservoirs, that is, as optimum environments for certain stages of a disease life cycle.

Malaria is a good example of the environmental limitations on the geographic distribution of a disease and therefore on its diffusion within a population or throughout the world. Malaria parasites cannot develop inside the mosquito in regions where daytime temperatures are too hot or nighttime temperatures too cold, illustrating the principle of the dependence of population distribution on optimum environment. These features of physical environment limit the geographic distribution of malaria to tropical and subtropical regions.

Bilharzia, a worm parasite that afflicts millions of people, again, in tropical and subtropical regions, has a life cycle that depends upon a small freshwater snail and man as primary and secondary hosts. It is not found among human beings outside of tropical regions because the snail host cannot live in cool climates. Thus, we recognize another example of the environmental limitation on the distribution of a particular organism. A human being infected with this flatworm can carry the adult stage of the parasite outside of the natural range of distribution, but cannot transmit it to other human beings because of the absence of a secondary host, the snail.

Although the general principle of disease diffusion depends on natural ecological principles, some diseases have been diffused artificially by man. Chestnut blight is a disease that was accidentally imported into the

United States from China, and therefore is an example of the extended range or distribution of a disease as a result of some artificial phenomenon. This disease struck the large and beautiful chestnut trees of the Appalachian forests and all but exterminated them. It is an example of the threat to the natural ecology of a region that often results from the introduction of some non-native species and just one of the innumerable cases where the local ecology has been drastically altered by an introduced species. But geographic extension, by diffusion, into unnatural areas is the case in point here. Only if a disease meets with favorable conditions can it endure. The case of chestnut blight stands in contrast to malaria, the former having had its geographic distribution extended by humans and the latter being confined by quite restrictive temperature requirements, despite the wide movement of one of the hosts, human beings.

Dutch elm disease, another example of artificial diffusion, was accidentally imported to the United States about one hundred years ago. It is a fungus spread by a bark beetle, which is not a host, as the malaria mosquito is, but is nevertheless a carrier. Theoretically, the distribution of this disease is confined to the range of the elm tree as a species and the range of the beetle that carries the disease. There are, however, other host trees such as ash, maple, and cottonwood, and other insect carriers as well, but most people do not take this into consideration when planning a Dutch elm disease control program.

The diffusion of Dutch elm disease was facilitated by the range of other host populations and by the transplantation of elm trees into regions where they were not native, hastening the diffusion of the disease. A further complication is that more than one kind of beetle can spread the disease and, unlike the malaria mosquito, the diffusion process has been much more rapid and extensive. More than half of the United States has been affected in less than one hundred years! As in all other disease transmission, the operative factor, diffusion, and the two conditions, density and distribution, are population characteristics that have played roles in the spread of Dutch elm disease.

It is strange, however, that people have not observed that elm trees, as a population, have survived a hundred years of Dutch elm disease. Millions of elm trees can be seen that are between one and two hundred years old. Inasmuch as the disease has been in America for one hundred years, it is obvious that the population dynamics of elm trees have maintained the species against this disease. We have mounted a crusade against the Dutch elm disease without fully using the knowledge we have of population dynamics and life history. A campaign against Dutch elm disease that is directed only at elm trees ignores the life history of the disease. This involves at least three other native tree species that are hosts, or nodes, in the diffusion network.

Dutch elm disease is only one of the many environmental hazards in the life history of elms. Mortality patterns in all trees reveal that the first logical step in shade-tree maintenance is continually to plant young trees in succession of age classes as replacements for those that are lost as a result of disease, storm, and other natural hazards.

Although the idea of disease prophylaxis in city shade trees at least keeps dead limbs from falling on our heads, it is purely a defensive tactic. As in so many situations, the American idea of a panacea or crusade often blinds us to using the great knowledge we have in a logical manner.

The Linkage of Ecosystem, Social System, and Political System

At this point we may pause to observe the trend in this book from ecosystems through human ecology to social systems. Human social systems are part of ecosystems because man is dependent on the feedback loops of biophysical and biochemical systems. Social and political systems are linked, and each of these is really only a subsystem in the ecosystem.

This trend in thinking recently has been brought into clearer focus by teams of researchers at the Massachusetts Institute of Technology. In particular, Jay W. Forrester has used the old biological idea of food webs and feedback loops, now called cybernetics, in the application of systems analysis to social problems. In his book *Urban Dynamics,* Forrester has called the American craving for stopgap solutions to problems *counter-intuitive behavior.* Only a dozen or so years ago, John K. Galbraith, who dubbed us "The Affluent Society," noted that we invariably use *conventional wisdom* to solve our problems. Conventional wisdom is intuitive. Sometimes we call it snap judgment. Intuitively, we leap at the first solution that comes to mind when we are confronted with a problem.

Day by day, we solve our smaller problems this way; but by systems analysis Forrester has shown that many of our social problems are only aggravated by this kind of problem-solving. He found that in the long run many of our "conventional solution" programs prove to be wrong. Forrester applied the term "counter intuitive behavior" to this human urge to solve problems by intuition instead of by systems analysis.

More recently, members of The Club of Rome have

become highly concerned about the impact of human population pressure on the world ecosystem. In 1970, this concern was formulated into a research program which they called the Project on the Predicament of Mankind. There is hope that the efforts of these scientists will reveal logical solutions to environmental and social problems that heretofore have been "solved" by "counter intuitive behavior," only to appear and reappear time and time again.

Whatever ideas these philanthropists generate, to translate them into public policy will require an enlightened population in a node and network pattern over which the diffusion process can function to spread an idea. They are dealing with the global ecosystem, and we can only hope that they will reveal solutions to the old problems of famine, disease, war, and pestilence —those perennial population problems that puzzled Malthus two hundred years ago.

X

Bedfellows: Starvation and Disease

A conscious crusade against hunger has gripped the world since World War II. Famine is a frightening thing and the thought of it raises compassion for the starved and hungry, even across national boundaries. Famine and disease seem to be bedfellows. Malthus wasn't quite sure which population control acted first to depress a population, and after all these years it still is not clear. Even India, the country we have regarded as the exemplar of starvation, suffers more from disease than directly from famine. Disease appears to be triggered by malnutrition, but hunger is not the whole story; even the well-fed suffer from disease.

Famines are nevertheless catastrophic, whenever they occur. The oldest record of famine is from the first book of the Bible, Genesis, in which the account of Joseph in Egypt is related. Joseph interpreted the dreams of

Pharaoh, predicting seven years of plenty and seven
years of famine. According to the Bible, his predictions
came true: after seven years of bumper crops there
was a severe crop shortage for seven years, and widespread famine followed in all of Egypt.

A great gap in the history of disease and starvation
exists between the times related in the Bible and the
Middle Ages. We commonly think of "biblical times"
as running into the Roman era, but the origins of the
biblical record are much more ancient than Roman
times. Famine harassed man even before biblical times,
but during the Roman conquests disease appears to
have been more prevalent than starvation as a population control. How else could the Roman legions have
marched from Italy to the British Isles? They must have
been supplied with food from the new lands they con-
quered as they marched.

About a thousand years later, great plagues caused
widespread and catastrophic death all over Europe.
There is much more mention in the history books of
death from disease than from starvation. Plagues are
more precisely documented; the years and locations
of plagues, as well as estimated number of deaths, are
often given. On the other hand, the record of death by
starvation is vague and spotty. In Western Europe,
where better records were kept, about 450 localized
famines are known to have occurred between the years
1000 and 1855 A.D., a period of 855 years. For so long
a period of time and such a large area, this really is
not an astounding record of starvation—unless you
happened to be part of the scene. It would appear from
the histories that these 450 famines were caused by
local crop failures, and that food was available on a
wider geographic scale; but because of the poor trans-

portation in those days, it could not be distributed adequately to stave off hunger in certain places. Whatever the consequences of famine were for the individual— in a given place at a given time there were whole families who might have starved—the general population growth curve in Europe does not indicate any stoppage in normal growth form. Records of censuses, of course, were not very good, and so speculation on this subject is probably quite fruitless.

Given its controlling influence on population, famine is rarely, if ever, an influence on the natural proclivity to reproduce; rather, its influence is indirect. It causes mortality of those individuals who have already been born, but does not reduce the inherent individual capacity to reproduce. To be sure, famine has been associated with war throughout the history of mankind. War disrupts the normal patterns of productivity and deters the cultivation of the land. Even worse, it disrupts communication and transportation, and this disruption can seal off areas which then suffer privation in the face of stupendous regional surpluses. Such was the case in Biafra in the late 1960s. Throughout the Biafran war, northern Nigeria produced vast quantities of meat, hides, peanuts, and other commodities for world export, while people in the next province starved!

We have had, in America, years of drum beating over the impending crisis of a food shortage. The theme has been that the population explosion will cause great and dire famine. Many popular books have carried this theme of imminent famine. One is entitled *Famine Nineteen Seventy-Five: America's Decision, Who Will Survive.*

American farmers have produced such an abundance of food during the past fifty years that they have caused a severe "farm problem" for every political administration since 1920. Among the many federal programs to counteract the superabundance of food produced in the United States, there have been such programs as the soil bank, land and water conservation laws, the agricultural stabilization and crop adjustment program, and a number of attempts to dispose of surplus food such as the school lunch and food stamp programs. Also there are the very controversial subsidies paid to certain owners of huge farms who are paid *not* to produce food or fibre on them.

There are four major eras in the history of American food production. During the presidency of Thomas Jefferson, the "agrarian polity" furthered the idea of national advancement and development based on a free market system of free and independent farmers. Less than a hundred years later, a second stage developed—the Homestead Act, passed in 1862 for the purpose of settling more people on the land, with the provision that they should "prove up" by making the land produce a living. These were national development programs based on the idea of agricultural development and sponsored by the federal government. A third stage implemented a reversal of the expansionist programs and can be datelined by the Forest Reserve Act of 1891; some restrictions were put upon the opening of new lands, but they were not totally restrictive. At about the same time the rather ambiguous movement of reclamation was innovated. The role of the Bureau of Reclamation was to make wet land dry and dry land wet! Nevertheless, the whole idea was to increase the area of agricultural land.

A fourth stage, covering a period of the past forty years, has been generally restrictive on agricultural production, even though the Bureau of Reclamation continues farm-land expansion. In 1933 the Agricultural Adjustment Administration, the Resettlement Administration, and the Soil Conservation Service programs were intended to alleviate the problems of overproduction of food. All of these programs survive in one form or another, even today, for the same reason; yet, the Bureau of Reclamation carries on with the "multiple use" projects, one major use being the irrigation of semiarid lands for agricultural development in the western states. These are controversial programs, "pork barrel" projects, intended to improve the local economy. Right now, in the 1970s, small farmers in the same places are forced out of business because of the disbalance between their costs of farming and their income from crops; we have at hand a "land reform," economically triggered, that displaces the farm family. This movement is one in which large manufacturing corporations, not primarily involved in agriculture, purchase the smaller farm units, amalgamate them, and engage in large-scale farming as a tax write-off.

This pattern of land reform is just the opposite of the one our government proposes for Latin America, although there the object is social and economic leveling of people, not crop reduction. In South America the trend has been to break up the *latifundia,* or large estates, and to apportion them out as *minifundia*—small, privately owned farms—so that the large landowning class is in some way expropriated. In theory this is supposed to redistribute the wealth so that the poorer people will have an opportunity to improve their economic status. Note well that this kind of foreign aid tends to make an agrarian nation out of those nations where such a policy is implemented. The effect is to impede the demographic transition that facilitated the improvement of both social and economic conditions in Scandinavia and the United States.

Is the Food Crisis Fact or Fallacy?

Famine versus food surplus is controversial. Whether one is optimistic or pessimistic depends on one's attitude on the subject. How does the function of the logistic curve fit in? The principles of biology reveal a virtual law of population growth—that population is limited by food supply. If the Earth is a finite environment with ceilings on food production, or on any vital commodity such as air, then we might well be apprehensive about the population explosion.

Much evidence for this viewpoint has been assembled in books such as *The Hungry Planet, Our Plundered Planet, World Without Hunger,* and *Population, Resources, Environment.* There are, however, other sources of information which lead to another point of view. The *Statesman's Yearbook,* which has listed national imports for nearly a century, atlases produced in different countries by independent fact-finding individuals, and world trade reports all show a constant increase in food production. The Department of Defense published a series of six books with the very catchy title *The Ecology of Malnutrition* about six different regions of the world. In no instance in these sources is an outright shortage of food production cited. Malnutrition, it must be remembered, is one thing, and food shortage is another. Malnutrition can occur in the midst of plenty; many American youth are poorly nourished because of their choice of diet—a cigarette and a coke!

Among the current controversies related to food production is the entry of Great Britain into the European Economic Community (EEC). Although the complexities of Britain's long delay are many, the fact that British entry threatens the national income of other food-producing countries is the heart of the problem. New Zealand and Australia are overproducing food and fibre and therefore depend on Britain as an outlet for this surplus, a situation with the same characteristics as our own farm problem—overproduction and superabundance that constitute a disbalance in the domestic economy. We have long fought this disbalance with our idea of "parity," the idea of giving the farmer a "par" chance of earning a reasonable profit for his effort and cost of production.

How old is this problem of international-food-exchange competition and production control? The Hanseatic League traded in foodstuffs in northern Europe from the late 1200s until about 1400. The Hansa, a guild of merchants operating from north German cities and trading with Polish, Swedish, Norwegian, Scottish, and other North Sea ports, traded in butter, salt fish, wine, furs, and prepared meats; the markets were controlled by boycott and monopoly. Perhaps this was the forerunner of the EEC idea!

In about 1400, England passed the "corn laws," relating to all grains, and some other foodstuffs as well, which were to protect the English farmer from the lower prices of food produced in other countries, such as France, Belgium, and Holland. Corn laws were passed and repealed several times between 1400 and 1600. New laws of the same kind were enacted to protect the English farmer from the competition of farmers on the Continent just after Napoleon's defeat. England has had this problem for nearly 600 years!

All African nations are agrarian, with, of course, products quite different from those of northern continents. A very different environment, tropical and subtropical, is ideal for world-popular commodities such as coffee and tea, which cannot be raised in northern population centers where demand is great. And yet these African nations remain as underdeveloped countries—agrarian—unable to make the demographic transition! Uganda, Kenya, and Tanzania compete with each other, as well as with Colombia, Brazil, and Honduras, for the world coffee market and other farm commodities (Figure 25).

Great surpluses of coffee and other commodities are produced and have been produced for years in these countries. Coffee is now cheaper in America than it was twenty years ago, when Brazil and other coffee growers burned tons of beans to raise the price! The history of international trade agreements is long and complicated, but the facts remain that for years these countries have produced surpluses of food, food for export. If these people are hungry, would they not produce something to eat first, before producing for export? Could they continue to work the fields, expend energy, without food? These are some simple questions that have not been answered by some rather high governmental committees.

In Africa, south of Sahara, domestic cattle herds are estimated at about 140 million head, in addition to an estimated several hundred million goats and sheep. The amount of meat on the hoof in Black Africa is stupendous. Some people argue that it is "poor quality"; but by what standards? It is still life-giving protein.

Many African tribes have social customs by which a man can trade animals for a wife, and this encourages keeping huge herds. Because polygamy is practiced,

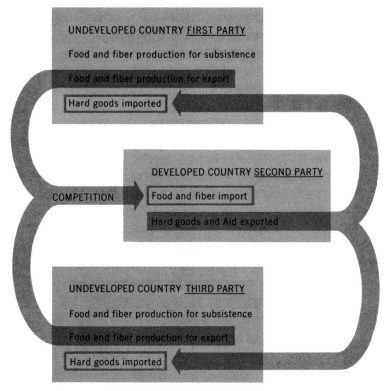

UNDEVELOPED COUNTRY <u>FIRST PARTY</u>

Food and fiber production for subsistence

Food and fiber production for export

Hard goods imported

DEVELOPED COUNTRY <u>SECOND PARTY</u>

Food and fiber import

Hard goods and Aid exported

COMPETITION

UNDEVELOPED COUNTRY <u>THIRD PARTY</u>

Food and fiber production for subsistence

Food and fiber production for export

Hard goods imported

Figure 25. Foreign aid tends to keep the underdeveloped countries in the position of agrarian states whose populations cannot make the demographic transition. Most UDCs produce large quantities of food and fibre which are in surplus supply, competing with each other for the world market, and thus are in an unfavorable bargaining position with the developed countries.

cattle have value for acquiring several wives. To change these ancient customs is more than can be expected; but it is mere folly to try to improve the quality of meat when the important point is the value of a live cow in trade for a woman.

In the province of Kano (about the size of New

Jersey) in northern Nigeria, there are seven leather tanneries, which have been operating for more than half a century. These tanneries process the skins and hides of an average of five thousand sheep and goats, killed each week, year in and year out, in one province! Where do the hides come from? Off of a carcass of meat that weighs about fifty pounds. That is a lot of shish-kabob!

The indigenous people of tropical and subtropical regions do not require the same diet that people of colder climates require. Life styles are different. In addition to the warmer clime, the pace of life is slower, and less food is needed. Even if more protein were needed, these people are clever at snaring the millions of small wild animals of the cow family, such as duikers and reed buck, so there is no need for them to go hungry in the land we have for so long described as teeming with "jungle life."

Hunger, where it exists, is in the cities where food is not produced, and where the unemployed, downtrod-den, and displaced people congregate. Why should people crowd into the cities where it is likely that they will only compound their own problems? It is because overproduction on farms forces a world-wide rural-to-urban migration. This reflects a universal surplus of farm products. If we were to travel the twenty-one cities of a million or more people in South America, visit the markets, and walk the streets, we could see the vast quantities of food spread out on the asphalt and con-crete of the city markets and streets. It comes from the hinterland, fresh every day and is more than what is necessary to meet the alimentary needs of the people who raise it. Daily, the people from the farms carry their produce into the city and each night return to their small farms.

Holland, one of the most densely inhabited countries in the world, produces food for export, while it also imports coffee, a food; cocoa, a food; and tea. The import of these foodstuffs is misleading. Unless the kind of commodity is known, the impression is given that Holland must import food to survive. Even though it would "go hard" on a Dutchman not to have his coffee or chocolate, he would definitely survive the ordeal. It is perhaps accurate to say that no nation is self-sufficient in food production by today's standard of living. Like Hollanders, the people of many nations have become accustomed to foods that cannot be produced in their particular climatic region, thus making international food trade desirable. Importation does not, however, necessarily mean shortage.

There is no way to enunciate with finality that we are short, or that we are not short, of food. Food production is changeable, with the seasons, and year to year. Good and bad years are really population problems: problems of insect populations, of plant disease populations, and of weather, the environmental factor having so much to do with all natural population dynamics.

There are other factors that can shed light on the famine scare. The trend in rural-to-urban migration of people is some indication of food-production trends. That vastly improved technology makes it possible to grow more food on less land is a cliché. So it follows that land put into retirement in the soil-bank program, for example, or under an erosion-control program, could as well, by the same improved technologies, produce much more food if needed. Obviously, it must not be needed. Farm families are consequently displaced by overpopulation and must move to cities in the hope of finding work.

Surveys of arable land not now under agriculture have been made in every part of the world, mostly by the Food and Agriculture Organization (FAO). It has not operated in China, of course, because China was not a member of the United Nations until 1971. Chinese geographers have reported, however, on the arable lands of northern China, which, although not as productive as those already under agriculture, are more extensive.

We too often equate productivity with "bushels of corn." Land which will not produce corn is put down as being "poor," even though it is not poor for everything and might be put to valuable use. The great forests of Canada produce a greater bulk of fibre than corn fields do. While we do not yet eat tree fibre, this is an illustration that diversity of environment and diversity of production are not usually taken into consideration when the word "productivity" is used. The great unused, but arable, lands of China can produce meat, hides, and wool when needed. In fact, they do so now and have for thousands of years. Some millions of people have made a living on these lands over a period of many centuries. India has almost as much arable land not yet under cultivation as it has under cultivation. Some experts believe that these lands are potentially high producers of manioc, otherwise known as cassava, or tapioca. According to the *Statesman's Yearbook*, 1971, there is more unused fallow land in India (58,490 square miles) than irrigated land (55,682 square miles), and nearly as much more as the total of both in arable land not yet under cultivation (97,720 square miles).

There are about 180 million cattle in India, nearly twice as many as in the United States, and in addition there are about 52 million working buffaloes, 40 million sheep, more than 60 million goats, and more than 120 million meat-producing chickens, ducks, and geese—if the people should want to eat them.

For many, many years the main export from India has been tea. In addition, coffee, sugar, jute, fish, spices, tobacco, peanuts, cotton, leather and skins (from the backs of meat-producing animals), vegetable oils, and a few other edible products have been exported in considerable quantity for a long time. All in all, the actual and potential productivity of food in India is more than adequate to feed the population—if it could be equitably distributed. The potential productivity is not even touched, and in most parts, as everyone knows, meat-producing animals are sacred, and therefore not killed, or even prevented from trampling and ruining crops. Remembering the food chain, we must also stop to consider the tremendous amount of green food that millions of beasts have to eat to stay alive, which could be converted to grain crops for direct human use.

While many kinds of land reform and agrarian reform have been tried in India, the matter of substitution of products has not been utilized to any degree. Various five-year plans have mainly been methods to redistribute land so that there would be a larger number of people on the land. This is a basic error—the same one that keeps Africa from making the demographic transition, a barrier that keeps them agrarian states instead of industrial states. An industrial state needs farmers, to be sure, but the agricultural dominance of exports shows India to be an agricultural state, not an industrial one, and if a nation can export food it should allocate this export to feeding its sons and daughters before it is exported. Other solutions to the so-called food problem are needed. The ones used heretofore have not worked.

The general condition in South America is the same.

Most likely our morning coffee comes from Colombia, Brazil, or Honduras. These three and several other Latin American countries compete with each other, as well as with African nations, for the coffee market. Argentina is the world's second largest exporter of meat, led by the very small country of Denmark, which produces the astounding amount of 807,000 metric tons of pork and bacon per year, much of it for export. These countries compete with New Zealand and Australia, as well, for the world meat market.

The population of most countries of the "non-Western" world—that is, Africa, Asia, and South America—is composed of about sixty percent farm workers, and a considerable proportion is not fully employed. Compared to the United States, with less than five percent of the population engaged in agricultural production, these countries are agrarian and in the same economic posture as our own farm community: in need of a reform policy that has not yet been devised. Our foreign aid programs and "land reforms" are failures. In the face of a generally oversupplied world food market, the bulk of the farm population is ensnared in a mesh of partial employment from which it cannot extricate itself.

The very thing that causes the unemployment of farmers also causes their hunger. Overproduction of farm commodities puts people out of work, forces a rural-to-urban drift of migrant population, and dares them to earn enough to buy the product of the land they left—food. Workers in industrialized countries, with their hard-goods production in full swing, can earn enough margin of profit over living costs to be taxable and therefore pay subsidies to farmers not to farm. In this sense, some countries are overdeveloped and have generated an unbalanced agricultural-industrial system. Instead of farming with horses that eat oats, they farm with tractors that burn fossil fuels, and instead of diversified agriculture with a patchwork of diversified crops, itself a hedge against natural hazards, great monocultures are created. These, in the form of vast single-crop fields, such as wheat and corn, are highly vulnerable to natural hazards. The huge fields themselves provide an optimum environment for a population explosion of such infestations as wheat smuts and corn borers. So, the farmer must employ chemical warfare—the crop duster, who compounds the problems in a series of interrelated events much like a food chain. For instance, to spread the DDT requires a flying machine that devours fossil fuel and leaves a residue of exhaust in the process. There is more to the picture than that, of course, but perhaps it is enough to convey the idea.

Our technology in food production is absurdly ahead of need. Announcements of new ways to make odd things palatable, such as sawdust to be fed to cows, keep people believing in the food-shortage fallacy. Synthetic foods are not made of "free" products. The base of any material is other material. The first law of thermodynamics, paraphrased, says that matter can neither be created nor destroyed. Synthetic foods must be created from other materials.

Food technology has created a bigger and better apple. Processing and packaging have done much to minimize waste and to influence consumer attitude. Cottonseed cake, peanut cake, and sunflower cooking oil are all products derived from surplus commodities; the puzzle is what to do with them. Find an attractive way to process and package them, and maybe the

public will buy them! Merchandising is the trick in
surplus commodity disposal: if you have too many
apples, make applesauce; save the housewife the trouble
of preparing it, and maybe some of the excess supply
can be marketed as a disguised food. Apples and applesauce are normal, natural commodities, but when the
same idea is applied to fish-meal bread, the impression
that people often get is that the "new source" of food
indicates a shortage of old sources. Probably the most
common example of the merchandising gimmick to sell
surplus food is the cereal box toy, as well as the various
ways that cereals are prepared to appeal to the child
consumer. Such processing and toy premiums increase
the apparent cost of food, which deceives the housewife
into thinking that the farmer charges too much for the
food he produces.

The other side of the cereal picture is the feed-grain
program. This is a program under which the federal
government pays farmers not to plant feed grains, the
same grains used in making breakfast cereals. In 1966
more than 1 billion, 250 million dollars were paid to
farmers to stop the overproduction of feed grains,
which are used to feed beef cattle and all other meat-
producing animals. This cost paid by the taxpayer is
part of the cost of the food we eat but is so well hidden
that we do not immediately see these production-con-
trol costs as part of the overall market basket food cost.
Six such programs are in effect in the United States
to prevent the overproduction of livestock feed crops,
which could also be made into cereal and bread—if
we needed them.

At each trophic level in the food chain, there is a
loss of efficiency. Recognizing this wastefulness, nu-
trition experts use the idea that if a level in the food

chain can be skipped, more top-level consumers can be supported by the basic green plant level in the food chain. In other words, to feed grain to a chicken and then eat the chicken is more wasteful than for people to eat the grain as cereal themselves, skipping the chicken-feeding stage. Borgstrom, in *The Hungry Planet*, has termed this idea "population equivalent." Therefore, as Borgstrom puts it, ". . . we should think in terms of population equivalents. . . ." He has calculated that the livestock in the United States consumes protein that is equivalent to the needs of 1 billion, 300 million human beings.

In addition to the close calculations of nutritionists, the work of the food chemists to find new sources and of the processing and packaging experts to make odd things look and taste good, a great effort has been put into research on the agricultural use of wastelands, such as deserts and salt-marsh areas. Food production in regions where plant requirements are not available within the natural limits of the ecosystem will be very costly. Cost, however, is relevant to need; if they need it badly enough, people can find ways to do something. Dollar cost is never ultimately prohibitive, but ecological cost can be.

It is a delusion to believe that food can be synthesized from nothing. All production processes require materials. Projected dollar costs tending to show the increased cost of producing new foods or finding new food sources are meaningless. If food is required, it is axiomatic that cost-benefit ratios will be equated, or else it won't matter; war, starvation, or production costs will equate with the population in accordance with the inexorable logistic curve. If this is the human choice, costs to the ecosystem might be bitter for man, regardless of dollar cost.

Epilogue on the Malthusian Bedfellows of Human Ecology

Famine, food, farms, factories, poverty, people, pollution, ghettos, disease, and illiteracy are all subsystems of a greater ecosystem. The problems of one are linked to the problems of all others. Food surpluses can cause famine. Food surpluses, or overproduction, cause rural-urban drift; urban drift causes unemployment; unemployment causes poverty; poverty causes ghettos. It would seem that we really haven't learned much since Malthus linked population problems with poverty, war, disease, starvation, vice, and crime.

The long crusade against famine—to grow more food —was aimed in the wrong direction. Or so it might appear were we to apply systems analysis to this very old problem. How else could world population have increased exponentially? And hunger with it? And poverty? The many facets of the hunger crusade have been diversionary and counter to the concept of ecosystems.

We can only hope that with world systems analysis we can evade the conventional wisdom, sidestep the counter intuitive solution, and tackle the problems of human ecology with modern systems analysis translated into a national population policy. We should tackle the *cause* of famine and poverty, not the result.

XI

Politics and Population Theory

Today's political arena seems to be a melting pot of social, economic, and ecological ideas. If it is the forum where human destiny is cast, then we should recognize a natural relationship between human ecology and politics. We have already discussed, in Chapter IX, the way that biotic, social, economic, and political subsystems are linked and webbed in the larger ecosystem of which man is part. In this chapter we shall deal in theory as a means of bridging the foregoing chapters with the closing chapters on the national estate and optimum population.

About the time of the American Revolution, nearly two hundred years ago, an Englishman, Thomas Malthus, propounded a theory of population that still seems to be the most popular and perhaps the most profound of a large number of such theories. Earlier in this book there have been several references to Malthus's ideas, but now they will be considered in greater detail and in connection with their political implications. The Malthusian theory seems to have covered the political and the economic fundamentals of the population mechanism rather well, and—in a way—it covered the ecological concept of the population mechanism, also. Basically, even wild animal foraging is economic. An elephant, for example, cannot feed efficiently in short grass plains. Its forage requirement, about four hundred pounds a day for a middle-sized elephant, is greater than the yield of a short grass prairie in relation to the energy that the elephant would have to expend in order to get that much food from it. Optimum environment for elephants is tall grass savanna, where foraging efficiency is greater for them. Thus, economics pervades ecology. In fact, the Greek root of these two words, literally translated, means "of the house." *Economics* refers to the "numbers" of the house, or housekeeping, while *ecology* refers to the "logic," or science, of the house.

Malthus's theory begins with a statement based on the idea of the population growth curve, the principle of exponential growth: there is a tendency for population to increase in a geometric ratio, that is, by multiplication (two times two is four, times two is eight, times two is sixteen, and so forth). On the other hand, Malthus believed that the ability of an environment to feed or support a population increased only in an arithmetic way: two plus two is four, plus two is six, plus two is eight, and so on.

When the operation of these two ideas is compared, it seems apparent that the growth of population soon outruns the ability of the environment to feed it. The comparison of the two ideas looks like this:

	Time				
	1	2	3	4	5
Population growth (by multiplication)	2	4	8	16	32
Food production (by addition)	2	4	6	8	10

This scale shows that everything is nicely balanced between population and environment up to the second time period, and then population begins to increase faster than food can be produced. This is quite in keeping with the idea of the logistic curve, but the actual quantitative characteristics are not quite as represented. Disease appears to be the first factor to limit a population that becomes too dense. Malthus recognized the role of disease in controlling populations but did not put it foremost as a control factor. In fact, he listed starvation, disease, war, and vice as population controls in that order. Really, all he seems to have missed in noting the depressant factors of population increase is accident. To quite an extent, accident seems to have become a major population control among human populations; it even outran Viet Nam war casualties in the American population.

A Scottish economist by the name of Adam Smith, whose ideas of population were more closely related to economic factors, lived at the same time as Malthus. Publishing a book in 1776 called *Inquiry into the Nature and Causes of the Wealth of Nations,* Smith propounded the idea that labor is more important as a factor of economics than land or capital is. The relation of labor to population is fairly obvious, an idea not restricted solely to Adam Smith and Thomas Malthus.

Mathematicians in France, Italy, and Germany were interested in population dynamics for some time before Smith and Malthus were born. But, because of the relevance of Malthus's theory to human problems, it is presented first here as a background for deeper consideration.

Machiavelli, an Italian statesman who lived in the era of the Renaissance from 1469 to 1527, was a politician concerned with population from the point of view of maintaining an army, rather than from an economic or purely political point of view. His name is associated with unscrupulous practices and principles designed to maintain the authority of the ruling class, for which a subservient army was required. Machiavelli therefore studied populations from the viewpoint of maintaining a supply of adult males for an army.

Broadly speaking, two general schools of population have been recognized: the mercantilist school and the Marxist school. Although it is risky to try to boil these philosophies down to a few words, it will suffice to say that the mercantilist school is interested in using people as a labor pool to control wages, production, and other business relationships, whereas the Marxist school is interested in the opposite attack, using economics as a means of managing population problems. Foremost in the Marxist scheme is the idea of redistribution of wealth, that is, the leveling of wealth, or wages, leaving the population to adjust to economics.

Biologists have become interested in population dynamics since economists and politicians have, and in a way biologists have provided a new school of population theory, the ecological. From this point of view, the other two schools could be put together as "economic" and stand in contrast to the "ecological." Probably some people would object to this simplification, but there is

really no basic difference between economic and ecological theories of population dynamics. All population
theories obviously must deal with populations; all populations are biological in nature; all populations react
to economics, whether primitive subsistence economics
or sophisticated financial economics.

In quite a useful way, an American economist named
William Graham Sumner modified the Malthusian
Principle; his modification, called the Neo-Malthusian
Principle, covers the subject more broadly in both its
ecological and economic aspects. Sumner proposed an
algebraic form of the Malthusian population theory:

$$P = \frac{L \times T}{S}$$

In this equation, P stands for population, L for land
(meaning all natural resources from air to zinc), T for
technology, or culture, and S for standard of living. One
cannot expect to plug in quantitative values for this
equation and have it work, but the philosophy of the
idea seems rather solid, no matter which way one cuts
it. If the Earth, L, is a closed ecological system, and P
is increased, then T must be increased to maintain a
constant S; that is, to maintain a certain standard of
living. These values can be transposed in accordance
with the laws of algebra, such as:

$$S = \frac{L \times T}{P}$$

This shows that to keep the standard of living up, with
a limited L, if P is increased then T must be increased.
Otherwise, more people, P, would be cutting the Earth
pie into more pieces, and S, each piece of pie, would

be thinner. On the other hand, if L were doubled and P and T kept constant, the pie S could be twice as big. It all agrees with the idea of the logistic curve, too. The very nature of ecosystems is both ecological and economic.

There is a thin line of distinction between theory and policy in dealing with human populations. Much of the theory has developed out of the need for some kind of population policy—to control labor, to raise an army, or for taxation. Ecological population theories include the theory of growth form (exponential growth), the theory of the logistic curve, and the theory of dynamic equilibrium. Economic, like ecological, theories are really few in fundamental form but have a number of variations: the Malthusian principle, the theory of demographic transition, and the Sumner Neo-Malthusian theory are examples.

Any theory must be in accord with the laws upon which the theory is based; a theory cannot contradict laws lest the laws be found erroneous. For example, John Maynard Keynes, in 1930, propounded a theory of optimum population. While this theory was essentially an economic theory, it dealt with people, biological subjects, and axiomatically had to conform to the biological principles of population dynamics, as well as to the principles of economics. Keynes's theory of optimum population was derived essentially from a rediscovery of the Malthusian theory and a growing apprehension about overpopulation, with consequential food shortages and unemployment. Some of the earmarks of Sumner's theory are also seen, serving to illustrate the tight interrelationship of all theories. A close look will reveal the play of the three factors of population dynamics—birth, death, and migration—

without which there would be no population, and no theory in the first place.

Population theories apparently interested people long ago. Ancient histories reveal that Aristotle and Plato were concerned about the demography of the Greek city states. Both of these philosophers are believed to have considered the idea of optimum population, their goal apparently not so much limited to determining how big or how small a population should be from the purely economic point of view, but they also studied how to achieve a high quality of life for each person. They recognized that there is a minimum level below which a population can not be economically self-sufficient. In view of present interest in population-efficiency levels, it is amusing to discover that Plato believed the optimum population to be 5,040 citizens,* the number most likely to be most useful to most cities. A population of this size was supposed to provide ". . . numbers for war and peace, and for all contracts and dealings including taxes and division of land." This needs a lot of explanation. A "state" with this many "citizens" had an actual population of about 60,000, with densities that varied from seventy-five to three hundred per square mile. Some historians believe that the Greeks did not live well at this density, and that the too-dense population might have been the reason why the question of optimum population occurred to some Greek philosopher-statesmen.

Even earlier, the Chinese philosopher Confucius was also said to have concerned himself with theories of

---

* Modern estimates by Zero Population Growth show optimum population for cities at about 100,000 people. This is close to the Greek idea of 5,040 citizens, a "citizen" being an adult male freeholder. Thus, women, children, slaves, and soldiers were not counted.

optimum population. Widespread famine led to speculation about what limits there might be to the efficiency
of a particular population from the standpoint of productivity to avoid famine. The Chinese concept of
optimum population appears to have been related to
the optimum number of farmers in the nation. This, of
course, leaves one wondering about the number of
soldiers, politicians, merchants, and craftsmen. At any
rate, the Chinese in the time of Confucius appear at
least to have considered the ideal ratio between land
space and population. The logic of this idea seems to
apply even today.

Population Theories and Labor

Most theories of population have to do with labor, one
way or another. Concern over productivity seems to be
a natural proclivity of government; consequently, much
theorizing about productivity has stepped over into the
realm of population theory. After all, production requires labor, and labor is either human or animal effort.
   Perhaps many of us, especially leaders in government, think of labor as something separate from people,
as if it were a big machine, rather than as so many
pairs of hands. Manpower is, of course, the labor force,
and the labor force is the proportion of a population
that can support itself and all the "drags." There have
been a large number of theories about the way a labor
force functions, or should function, to achieve certain
ends. During the seventeenth century, from about 1650
to 1700, there was a considerable effort on the part of
royal governments to aggrandize nations by clever
manipulation of the labor force. At this time, the so-
called mercantilist and cameralist (people of the "cham-

bers," or rooms of government and business) schools of political economy proposed a surprising number of theories to increase the wealth of the state, which we now call the gross national product (GNP).

In general, the idea was to promote a large and growing population in order to enhance the economic, political, and military advantages of the state. Behind the idea of a large and growing population was the belief that growth would create an excess of national income over the cost of production in wages, which would leave a large excess that could be taxed, thus increasing the private incomes and political power of the cameralists. Specifically, the theory supposed that growth of population would produce a large labor force, which, following the law of supply and demand, would depress hourly wages. In turn, that would create an incentive to work longer hours and thus widen the margin between national income and wage costs. It does not take a very close comparison of this theory with today's conditions to see what might happen—and in places, what is happening.

Several of the more modern nations, such as those of Fenno-Scandinavia, have gone through these experiences and have matured demographically and politically and as a result have achieved a balance between the labor force and the dependent groups. This specific demographic function is called "replacement ratio," the proportion of old people retiring from the labor force to young people entering the labor force. A nice comfortable balance of these two population segments is called a *one-to-one replacement ratio* with the result that there is no unemployment.

Among the economic theories of population, there are two or three that have an amusing twist, but when we think about the relevance of these theories today, the amusement turns to chagrin. In 1755 an Irish economist, Cantillon, theorized that if wage-costs of producing food become inefficient, a country should import food and export manufactures. Britain has been doing this for a long time. During the French Revolution, Britain had to pass "corn laws," which really applied to all farm products, to protect the British farmer from the lower, competing prices of farm goods from the European continent.

In 1832, about twenty years after the most restrictive corn laws were passed, Thomas Chalmers wrote about population theory without knowing, apparently, about the logistic curve. He believed that if the market were overladen with corn, there would be a population increase. He deduced that in this way a surplus of corn would create a market for itself. This is quite contrary to the idea of the logistic curve and very much like the idea of sending wheat to India to alleviate a population problem. Theory is often useful, but sometimes blinding. Even today the same problems prevail—in Britain and in Europe, primarily—because all of those countries produce such a surplus of food that it is difficult to keep their farmers in business. That has been the basic reason why Great Britain did not join the European Economic Community until 1971.

There were other theories besides the mercantilist or cameralist ones that are more palatable today. Ludwig Brentano, observing that an increase of income enhances culture, assumed that if a wage earner were to make more money than is needed to subsist, he would improve his intellectual status, which, in turn, would lead to a limitation of the size of family, and thus, indirectly, to a stabilized population. When we compare the demographic transition to this idea, it seems almost a truism, at least as a long-term generality. A corollary to this was observed by the British economist, Patten, who thought that civilization is the principle antagonist to the law of diminishing returns, meaning more or less that the whole idea of civilization is to make life easier. Productive power, then, is the principle antagonist to population increase. The ideas of these two men, Brentano and Patten, are very nearly identical. Several nations of the modern world are quite good examples of the validity of these "laws" of population.

Other variations of these ideas have been propounded. One theory of population that is quite like those stated above holds that every stage of advancing civilization diverts the people anew from their primitive appetites and passions. The result is supposed to be a voluntary limitation on procreation, and thus a limitation of population growth. Probably some readers will see a certain contradiction in the idea that civilization can be advanced and population limited at the same time, but various national population characteristics must be compared to national achievement to reach any sort of conclusion at all; later we will devote some consideration to this.

Whatever the economic, political, or social goals of population theory, biological theory is fundamental and inexorable. Economics, politics, and social dynamics can influence birth rate, migrations, and even death rates, but the irreversibility of birth and death and the mix of migration are so basic as to be more predictable than economic or social influences on the size and structure of population.

Population theories are useful and actually can be applied to our advantage—in insect control, for example, and in livestock production. Demographic principles of population growth also are used to plan construction for school buildings, to plan selling strategy (in market research), and to predict the outcome of political elections (by polls). Politicians might not be able to predict with much certainty the political attitudes of people, but automobile makers have a pretty good idea, premised on the idea of population growth form, of the future market for new cars.

Although in the long run population predictions are sometimes right, sometimes wrong, they are useful. Demographers have to work by hindsight, from records of the past, and cannot always outguess the public mood. Beneath all the theory, prediction, and guesswork, the factors of population dynamics—births, deaths, and migrations—often function quite independently of social and economic theory and in response to more basic human biological urges.

XII

The National Estate

A national estate is something like a farm populated by animals: it has boundaries, functions as a subsystem within an ecosystem, and is quite susceptible to the rule of the logistic curve. Every farmer knows that his farm can support only so many animals, and to manage it he must trade its products for others from different subsystems. Each nation in the world system—like a farm—has boundaries, has a population, produces and consumes goods, and finds international trade generally beneficial. But like a farm, some of the national estates of the world are beginning to find that they have production limits.

The national estate should be pictured as a bounded environment from which a population garners a living. Of course, the more conventional view is of a group of people with a common cultural heritage composed of language, ethnicity, customs, and life styles. But to complete the picture, environmental characteristics must also be considered. Thus, a national estate is a kind of ecological unit subject to the rules of ecology; it is an environmental unit with a dependent population.

We used to speak of the "have" and "have-not" nations in the world community. Since the forming of the United Nations in 1945 we have renamed these two groups the "developed" and the "underdeveloped" countries, usually abbreviated as "DC" and "UDC." Some of the so-called UDCs are very "rich" in natural resources and yet "poor" in the sense of social and political development. Some have an abundance of two of the factors of production, land and labor, but have not been able to muster the third factor, capital, also necessary for production. In some cases, the limiting factor appears to be the manner in which the population utilizes its national estate. Even though there is an abundance of labor and natural resources, the population has not, in many countries, utilized land and labor in the most advantageous fashion for the production of capital with which to diversify their social, political, and economic ventures. The worst overall problem is a demographic one. Most of the UDCs have an expansive type population structure, an age pyramid that is broad at the base, showing a very high birth rate and high death rate, resulting in a poor balance between the dependent segment of the population and the productive segment, the labor force. Therefore, much of the personal labor effort and resource drain go into the support of the unproductive segment of the population. What is needed is a demographic transition, not land reform and not a new agricultural technology, which seems to have been our mission as big brother in the community of nations.

Many of us in our conversation use the designations "DC" and "UDC" blithely in terms of natural resources, but the status of a nation's population in relation to its national estate is more important. The United Nations has listed criteria of national developement which are the basis for classifying countries as DC or UDC. Several of the criteria are:

DC	UDC
literacy, high rate	literacy, low rate
infant mortality, low rate	infant mortality, high rate
mean expectation of life, high	mean expectation of life, low
per capita income, high	per capita income, low
gross national product, high	gross national product, low

The demographic characteristics in DCs and UDCs, however, are not necessarily consistent with economic status in the family of nations. Even some developed countries have growing populations which produce a disadvantageous labor-force ratio, with consequential unemployment.

The international status of any nation is first a matter of "what you do with what you have." Even a small nation such as Denmark—just twice the size of New Jersey—with a land area of only sixteen thousand square miles, yet very much an agrarian nation, is among the five most prosperous sovereign nations in the world. The Danes manage their national estate in such a manner as to provide a very high standard of living for all, and moreover, have had very little concern over environmental stress. The national position of the Danish people as a DC is first and foremost a matter of managing a stabilized population. Nothing reveals the national character of a population more succinctly than an age pyramid. Of course, it does not reveal the mental

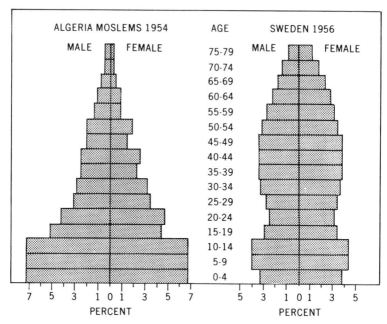

Figure 26. Developed countries and underdeveloped countries have different population structures. The better-organized countries have a favorable proportion of the labor force to the dependent group, and that reduces the burden of social costs on the productive segment of society. Therefore, literacy rates are higher and health conditions are better in DCs than in UDCs.

outlook of a population but sometimes even this can be detected from an age pyramid (Figure 26).

You may remember from Chapter II the three broad classes of population structure as indicated by age pyramids: constrictive, expansive, and stable. These three types actually found in the real world are quite convenient for classifying the broad demographic character of national populations. Remember that an age pyramid tells all about the dynamics of a population but says nothing of its size or density. On the other hand, the size of a population in relation to its national estate is a measure of environmental stress. Various nations with densities of between forty and one thousand per square mile provide a scale by which environmental stress can be estimated.

Nothing so vast as an environmental-stress scale can be reduced to a simple table, but at least the idea can be explored. The countries which are talking about pollution, that is, environmental stress, are those with relatively high densities regardless of demographic structure. On the other hand, those with disbalances of the labor force suffer unemployment, low per-capita income, and concomitant social and economic stresses, even in the presence of a bountiful national estate.

Holland, for example, has a distressing density that is poorly housed. The scarcity of land space is so acute that there are some thousands of families living in houseboats on seriously polluted canals, while at the same time, Holland can produce so much food that it exports relatively large quantities for so small a country. For a population of about 13 million people, there are about 4 million cows and 4 million hogs! Cows and hogs need standing room, not to speak of the space required to produce food for them. Still, in terms of health, education, and general welfare, the Dutch are quite well off. Although birth rate has been reasonably controlled, there has been a higher than average immigration into Holland from other countries, and it might be necessary to curtail immigration in order to prevent further distress of environment.

Bolivia is considered to be an underdeveloped country. With a vast wealth of strategic mineral resources, its national estate is capable of further development, but not without risk of environmental stress. Environmental alteration need not go so far as to create environmental stress, and yet resource development would necessitate some environmental alteration. Bolivia, with a high birth rate and high death rate, needs more than resource development to advance the populace to the status of a DC country. The population is of the expansive type, with a huge burden of dependent children. If the productive population were balanced with the dependent population, the resulting stabilization would do more to change Bolivia's status from UDC to DC than foreign aid or land reform could do.

A number of other UDC countries in South America have the same set of problems: they are distressed socially and economically, but not environmentally. Consequently, these countries provide some sort of a scale for a predictive model of environmental stress; their conditions give some clue to the prevention of environmental stress. On the other hand, the social and economic stress in these countries are related to population characteristics and the manner in which the population utilizes the national estate. While the most obvious factor preventing environmental stress is a low order of technology, the controlled application of technology is necessary to utilize the national estate in the best interests of the population. We are then faced with the decision at what level technology should be limited. When we consider the Neo-Malthusian equation, $P = \dfrac{L \times T}{S}$, the next factor that it would seem reasonable to manage is P. The population structures of these countries limit their economic and social development more than the generally low order of T, technology, does. It is useless to pour quantities of economic and technological aid into a national system without bringing about a demographic transition.

Nothing can do more to run a national estate into

bankruptcy than unlimited population growth. A sovereign nation with finite geo-political boundaries looks very much like a living model of the logistic curve. The smaller the geo-political unit, the closer it is to feeling the pinch of overgrowth. Although international trade can prolong the time before environmental stress catches up with "living too high off the hog," international trade is, in effect, an extension of geo-political boundaries and ultimately cannot evade the limitations of the global community of national estates.

It appears that the countries suffering from environmental stress are DCs and not UDCs. Countries in the DC category are quite generally overproduced in food stuffs and so there is no stress from that standpoint. The stresses in the DC countries are pollutions—water pollution, air pollution, and noise pollution, and, in some places, radiation pollution. These are the products of technology, the tools used to convert the resources of the national estate into a living standard for the population.

Earth Day 1970 for some people was the beginning of a reaction triggered by environmental stress. It is notably an American idea—we love a crusade! Why should such an idea have begun in America when certainly there are denser populations in the world? There is no nation with so many cars, telephones, bathtubs, electric razors, TV sets, or radios per capita in the whole world. This high "S" has required a high input of "T," causing environmental stress that is visible and resulting in an Earth Day crusade in response to that observation. The national estate is threatened!

Our attack has been a conventional one, like studying remedial reading instead of learning to read correctly in the first place. Launching attacks with the President's Council on Environmental Quality, pollution control commissions of all kinds, and crusades for natural environment, we are using "conventional wisdom" and treating the symptoms instead of the cause. Just as we ignore the principle of the logistic curve, we ignore the principles of demography: too many people with too high a proportion of dependent children. Intuitive solutions prove, in the long run, to be wrong.

As another example, we blame cars for most of the air pollution and set out, by technology, to reduce exhaust by fifty percent. This is meaningless if we add another one hundred percent of cars. The tactic of problem solution by technology is a consumptive one. Pollution-control processes require the consumption of materials and generate more environmental stress, consequently not only draining natural resources but also altering natural environment. Therefore, technology has a compound impact on environment.

Again, we encounter in the principle of the logistic curve the idea of limited environment. Instead of the crusade to treat the symptoms of the threat to the national estate, we should attack the cause: population pressure. Poised within the framework of the national estate is the balance between population pressure and environmental stress. The idea of *optimum population* seeks this balance.

An Approximation of Environmental Stress

A reciprocal ecological relationship, called the "green-grey ratio," exists between the areas covered by green vegetation and those which have been developed, such as large parking lots, apartment-house blocks, and factory space. The more asphalt and concrete we have, the less green cover, and vice versa. Green vegetation cycles oxygen, carbon, and hydrogen; asphalt does not. In fact, any area covered with asphalt or buildings is likely to operate in a negative way, that is, to produce exhaust fumes which reduce the availability of oxygen. Of course, it is apparent that as population increases so does the area of asphalt and concrete—and at the same time the demand for oxygen! Thus, a compounding stress is put on an environment by an increasing population.

Environmental stress and carrying capacity are borderline conditions. At the limit of carrying capacity, environment begins to show stress. We recall from the logistic curve that the carrying capacity of an environment is density independent, that is, it has a fixed limit, but here we see that environmental stress is density dependent. The more people in an area, the more stress they put on the environment. As a starting point for approximating environmental stress, we can compare national density with the proportion of a national estate which is in "green" and "grey" areas. Although we do not know how to measure environmental stress, what we do know provides a starting point for evaluating the carrying capacity–environmental stress relationship of a national estate. Each nation, of course, is a separate example, and a few for which best records are available as an illustration.

Sweden: area, 175,000 square miles; population, 8,000,000; density, 45 per square mile. Spatial allocation of the national estate:*
  55 percent Forest
  12 percent Tundra
  10 percent Water

---

* Approximations obtained from *Statesman's Year Book*.

12 percent Farm
11 percent City, urban, road
Green-grey ratio: 9 to 1—Balanced agricultural-industrial system.

Finland: Area, 130,000 square miles; population,
4,750,000; density, 35 per square mile.
Spatial allocation of the national estate:
70 percent Forest
10 percent Lake and bog
10 percent Farm
10 percent City, urban, road
Green-grey ratio: 9 to 1—balanced agricultural-industrial system.

Mexico: Area, 750,000 square miles; population,
45,000,000; density, 60 per square mile.
Distribution, vastly scattered except for
some large cities. No green-grey problem
because of general underdevelopment.
Exporter of food-fibre; an agrarian state.

Uganda: Area, 90,000 square miles; population,
8,500,000; density, 90 per square mile.
Distribution, vastly scattered except for
some large cities. No green-grey problem;
minimum heat-power-industrial pollution.
Exporter of food and fibre; an agrarian
state.

The United States is far from any critical condition
of overall environmental stress, but is the further
reduction of vegetative cover necessary to have a higher
standard of living? If we continue to grow, then, of
course, there must be further alteration of the green-
grey ratio to accommodate the national density in-

crease. In some localities environmental stress has already been observed, to which we have intuitively answered with the Green Belt movement. Again, we react intuitively! In the long run we could run out of green belt. Density and distribution of population are clues to carrying-capacity tolerances. Green-grey ratios reflect the pressure of population on the carrying capacity of the land: the more population, the less greenery!

Of all the assets of a national estate, the most important is people. The demographic characteristics of a national population are by far the most important component of the national estate. A high-quality population, tantamount to national prosperity, should have a high incidence of vitality, that is to say, a high level of national health. High literacy rate and a favorable balance of the labor force with the dependent groups are required to bring about such national characteristics, all of which are interacting and interdependent features of a national population. One feature feeds on another in a system of positive and negative feedback loops. People alone, however, are not the whole picture. People on the land and the relationship they feel with their land, their national estate, has much to do with national welfare. The trick seems to be to avoid environmental stress. If we may sum up the experience of neighbor nations, so far, it seems that the preservation of a favorable green-grey ratio is the best measure of a national land ethic and the way people feel about their environment.

XIII

Optimum Population: The Political Imperative

Signs of environmental stress show that, world-wide, national estates are in trouble. Just how serious the trouble is in various parts of the world, we really have not learned to measure. But we have perceived that environmental stress is reciprocally related to population pressure, that population pressure is a cause, and environmental stress a result. This cause-effect relationship reveals the key to balanced population in relation to carrying capacity. As we have seen, the carrying capacity of environment is limited; it seems obvious, therefore, that population must be limited. This ideal of a balance between population and environment has been called *optimum population.*

The concept of optimum population is not new. Like so many other ideas, probably as a result of famines, Chinese and Greek philosophers thought about optimum population two thousand years ago. Five hundred years before Christ was born, Confucius is said to have pictured the optimum population as being an "ideal proportion between land and population." In an agrarian society, that meant a sufficient proportion of the population engaged in agriculture to feed the whole population.

It is somewhat startling to discover that the Chinese philosophers of that day, visualizing what we are now calling a "population dispersal policy," or "new cities program," proposed that government should move people from areas of too-dense settlement to sparsely settled areas in order to maintain a favorable proportion engaged in farming. Chinese scholars also recognized the need for birth control to aid in the maintenance of optimum population. Famine brought out the cause-and-effect relationship: that the custom of early marriage produced more children and resulted in higher infant mortality.

Greek scholars seem to have been thinking the same thing at the same time in a different place, as translations of their ancient writings have shown. For one thing, Aristotle is said to have grasped the idea of personal development being related to optimum size of a population. It seems to have been recognized that "the good life" for the individual was tied in with a population of such size that it could efficiently provide all of the necessary services without overtaxing the capability of the land to support it. Aristotle wrote that the size of a population would have to be limited because land space could not be increased. He seems to have anticipated the idea of logistic growth! If property could not be expanded, civil strife would ensue; therefore, population would have to be limited. Plato and others recommended various methods of birth control as well as social pressure to increase population, if needed, to support the state in times of war. Thus, there is really quite a long record of the philosophy of optimum population.

A new concern about population is beginning to catch on. It started at the grass roots, as it should in a democracy, with such citizen-based movements as Zero Population Growth and Planned Parenthood. It has invaded religious principles in some quarters, but it has pervaded government only superficially through the Department of Health, Education and Welfare research on birth-control methods. We are still reluctant to allow government regulation of such personally delicate matters as who should or who should not have children. But, the government-sponsored research is aimed at providing a wider choice of contraceptive methods and therefore a wider spread of human choice in keeping with our ideals of self-determination.

So much for the "why" of optimum population; the "what" is really not very much different today than it was when the ancient Chinese and Greek philosophers thought about it. It is the same as Toynbee seeks— maximum welfare for the individual. Remembering that a population is composed of individuals, then what is good for the individual should come pretty close to being good for the population. However, since Aristotle's time a new concept of "quality of life" has been imposed on us by the environmental stresses generated by our level of technology: because of pollutions, we think of quality of life in terms of pure air, pure water, quiet, and the elusive "happiness." These are the environmental amenities that the new crusade is all about.

There is another dimension to quality of life—the vigor of a population. A population is composed of individuals, and without personal vigor, the population inclines to morbidity and consequently low productivity. National health, more important in matters of international respect than we have taken time to note, is contingent upon the personal health of its citizens. Quality of life enhances international respect. The prowess of less-populous nations like Finland, Norway, and Sweden at the Olympic games is an example of this point.

More formal dimensions of the quality of life are recognized by the United Nations and used to distinguish between developed and underdeveloped countries. Literacy is the key: a nation that can read (it is believed) can think. Usually, that nation also has a high health standard. When the chips are down, literacy is probably the first measure of the quality of life.* We in the United States, considering ourselves to be the most affluent of nations, have general and widespread stress in our schools and an alarming degree of illiteracy. Population growth—an expansive type of population—is the worst enemy of the American school system. At least two factors mitigate against a higher literacy rate: the reduced efficiency of crowded schoolrooms and the reluctance of taxpayers to meet the increased cost of schooling caused by overcrowding. This is purely a demographic problem, an unbalanced ratio between the labor force (which pays the bills) and the dependent children group.

The "where" of optimum population now encompasses the whole Earth ecosystem. Just as the necessity for international trade binds together the separate national estates, the whole world community is bound together in the ultimate need to maintain an optimum

---

* Finland, Norway, Sweden, and Denmark, for example, are said to be 100 percent literate.

environment. Pollutions are now global. We have not reckoned fully with atmospheric circulation, one of the most fundamental physical features of the world eco-system, and the global pattern of atmospheric circula-tion has spread radioactive fallout all over the world. Even if toxic levels are not yet serious, global spread shows, nevertheless, the potentiality of world-scale pollution hazards. Increased inputs of pollutants are proportional to increased population. There is no choice but to optimize population at a level where the amen-ities of a high quality of life can be maintained below the pollution level that global ecosystems can tolerate.

Necessity has always generated innovations in the minds of men, just as the stresses of environmental de-terioration are doing today. The ideas of new cities, rural redevelopment, and population dispersal have developed in response to population pressure. Con-sidering the node and network theory of geography and the exponential growth theory of biology, these solu-tions would transfer the problem of environmental stress to new places and defer them to new genera-tions. Like the diffusion of disease, environmental stress would be spread like cancer by the "new cities" and rural-redevelopment schemes—another example of counter-intuitive behavior!

Great holes in the canopy of population, such as the Sahara, have been considered for the settlement of excess population on the presumption that if oil could be found in the deserts, whole new populations could flourish there. This would put a new drain on the productive capacity of the arable part of the world, for the desert could not feed such a population. Ulti-mately, the result of an oil strike in the Sahara, or anywhere, would mean further atmospheric pollution everywhere. The global atmospheric circulation system would see to that!

Viewing such places as the Sahara as a bottomless pit where excess population may be dumped is to ignore the interdependency of all the regions of Earth in the global ecosystem. Because of basic human physical needs, which cannot be changed, places such as the Sahara, the Moon, and the Antarctic continent can support only what the arable part of the Earth can provide for, even if oil, diamonds, or the rarest of rare minerals are found there. In other words, the idea of "new cities" is limited by the environmental ceiling of the whole Earth ecosystem.

Finally, the "where" of optimum population can be examined in the light of old ideas about "have" and "have not" nations. Who "has" pollution, who "has not"? Pollution is, of course, becoming universal, but stresses are worse in some places than others. Generally speaking, the nations with the combined demographic characteristics of lowest density and lowest birth rate have minimum pollution, as well as a higher quality of life in terms of literacy and national health. Can these nations hang on until the rest of the world discovers the virtues of population control? In the world net of resource drain, they could be pulled under.

In short, this reveals the eventual need for a universal optimum-population policy. Concomitant with the idea of optimum population is optimum environment, minimum environmental stress, and world peace. Seeking these goals is a logical function of the United Nations as a whole, not just the Food and Agriculture Organization; according to the principle of the logistic curve, the FAO merely lifts the environmental ceiling so that more people can breed more people. Treating the symptoms instead of the cause is not the answer. This leads to the question, "Who?" Who should lead?

The Political Imperative

After World War II, the United Nations was set up as a world court to mediate the problems of the ever-tightening international neighborhood. Some nations are older and have been politically organized longer; some are further advanced technologically; and some are just plain bigger than others.

Developed countries will have to be the leaders in any policy of global ecosystem welfare, because only they are in the affluent position where such problems become visible. The underdeveloped countries are too busy trying to make ends meet to see their own gut-problems; they are too busy meeting the obvious problems of the day. They need help. The reason they should be helped is because they are part of a global picture that involves every nation in the world in an ecosystem that includes not only global environment, but economics and politics as well.

It follows that the principle of "enlightened self-interest" must be employed by the leader nations who have the wherewithal to do things. The Japanese and the Swedish people have already implemented population policies on an official political front. These movements toward population control came from the people. The same movement is coming from the people of the United States, but it is only beginning, and the rationale is not yet clear.

Sweden and Switzerland have been politically stable for a much longer time than any other nation. More than twenty years ago the women of Sweden organized a birth-control mission to India. Having discovered a better life for themselves and their children, as a result of birth control, the Women of Sweden, by their contributions, sent a mission to the Women of India to help them to attain the same kind of liberation. These are the courageous leaders, the people with the most to gain and the most to lose. Women! It is no mystery that the women of a nation that has been a leader in international law for more than five hundred years have courageously sought to aid the world's most distressed nation.

The United Nations is already organized for a job of this sort at the international level. The World Health Organization, especially, is equipped for such a job. The International Labor Organization has a deep interest in population problems, too. Possibly the simplest and strongest case for United Nations action on a global population policy is its responsibility to extend the doctrine of human rights throughout the world. The single most feasible step in this direction is population control to achieve a balanced, literate population structure.

At the national level, as has been noted, the leader nations have been those of low population density and stabilized populations, such as Sweden and Finland. Next in line, the developed countries with environmental problems should examine the logic of the environmental crusade. The shift should be made to a direct population-control policy instead of wasting money and manpower on environmental solutions that consume resources and frequently generate new pollutants.

The age pyramid of the United States population shows that the American people voluntarily controlled births during the depression of the 1930s. At that time, thoughts of "population explosion" or any other kind of population problems were not recognized by the American people. Rather, it was purely a personal response to economic conditions—hard times! This was all accomplished without any government action, too. This demographic phenomenon shows that change is possible and can be made quickly enough to be effective during a lifetime, and that people living now can, indeed, cause change.

If we review the development of human ecology, many would agree that the first criterion of optimum population could be derived from the long history of human experience. Man might devise a set of social principles such as the Mosaic Laws, or a constitution, but fall victim to the biological impulse of war because of population pressure on the national estate. National security can become a paranoic condition that diverts attention from intellectual alternatives to war. Population control could be the key to peaceful international coexistence and an alternative to power solutions.

We might also perceive a second set of optimum-population criteria derived from the United Nations criteria of national development. These identify with the individual and seem to be firm requisites of optimum population. The ability to read and the personal health of the individual (and health is dependent to a large extent on literacy) are criteria of the quality of life for the individual—hence for the population. Such criteria of optimum population are closely related to the demographic structure of a national population.

Other criteria that must be considered in devising a population policy must bridge the population/resource relationship. Feedback loops in ecosystems define the interdependence of all populations and a favorable environment; thus population in relation to the national estate has been called the *man-land ratio*. The necessity to manage the national estate, the environment from which the population derives its standard of living, is becoming increasingly obvious to more and more people. An optimum population is one that can live well within its national estate and in compliance with the biological principle of the logistic curve.

A National Policy for Optimum Population

The American crusades against environmental pollution are like a farmer who hauls manure away from a fenced-in field that is overpopulated with livestock. Pollution problem-solving requires the use of other resources; these, in turn, themselves need to be rescued from overuse.

There is no question that the national estate of the United States can sustain more people. The question is, is it necessary to do so? Could the United States be secure in all respects with a stable population, or even with a reduced population? Questions of national security, quality environment, and high living standards must be equated with the gross national product syndrome that grips us and results in the environmental stress that is reducing the quality of life for all.

The basic goal of the concept of optimum population is to balance the consumptive processes of life with the productive processes of nature. There is a geographic differential in carrying capacity of different parts of the Earth. There is a great contrast in the livability of the Sahara and of Iowa. National densities of population vary with the productivity of national estates. It would seem that the most obvious measures of overpopulation that we now have are environmental stress in terms of pollution and economic stress in terms of how difficult it is for people to pay the bills for raising and educating their children.

Figuring out the "how" is not a one-man show any-more than any other business deal is. To do so will require that the people perceive the nature of the prob-lems and direct their governments to seek solutions. Or, we might hope for a courageous and enlightened leadership to spell out the criteria of optimum popula-tion. Leadership should come from government, mo-tivated by an enlightened populace. The structural organization already exists for achieving the status of an optimum national estate in a community of sover-eign nations, but it is going in the wrong direction. The leadership has not reckoned with the logistic curve or with demographic alternatives to the Malthusian di-lemma: war, pollution, starvation, and vice. In a repre-sentative government, the people should stand the responsibility of change—especially change in attitude toward growth, from the GNP syndrome to growth in personal development. We are all part of a big machine. The machine has given us problems, but it has also given us the ability to recognize them. More people are seeing the problems more clearly now than ever before. More people are asking the question, "Is bigger better?"

The momentum of old established ways is hard to overcome, but we now understand the role of people in the world ecosystem better than ever before. More people now have more choice. Changing life styles are now recognized. The "new" equation is the population-environment equation, the oldest equation man has ever tried to solve. The choice of population control is the only solution to this equation. It is the political impera-tive for worldwide human welfare.

Appendix A

Construction of Life Tables

Life tables are really very simple. The first look at one may seem as gloomy as the darkest hour of night, just before dawn. After a little study, the life table begins to radiate the most exciting part of population ecology, like a bright sunburst that evaporates the darkness. As its title implies, the life table reveals the vitality and dynamic quality of life. Unhappily, the life table also depicts the tragedy that befalls all populations. For more background on sources of data and use of life tables, please refer to Chapter IV. For more detail on construction and interpretation of life tables, read on. Compare the relationship and similarity of Table 1A to Table 3, Chapter IV.

The first step in preparing a life table is to find what proportion each group is to the whole population, which equals, of course, one hundred percent. In Table 1A, for example, the proportion of five deaths in the first age class is one-third of the total number of deaths (fifteen),

TABLE 1A
Model of Basic Life Table Calculations

Age Class	Number That Died in Age Class x	Proportion of Deaths During Age x Per 1,000 Born	Number of Survivors at Beginning of Age x Per 1,000 Born
x	d_x	d'_x	l_x
1	5	333*	1,000*
2	4	267	667
3	3	201	400
4	2	133	199
5	1	66	66
	15	1,000	0

* These figures show that of every 1,000 living things born at the beginning of the time interval, 333 die during the first time interval, shown in the x column.

the same ratio as 333 is to 1,000. In the second age class, four out of the base of fifteen is proportionally the same as 267 out of 1,000 and so on down the table to the last age group, where we find one out of fifteen to be proportional to sixty-six out of 1,000. A simple check of the arithmetic is to add up the d'_x numbers (third column). If they come out to 1,000 all is well, for the "1,000" is equivalent to one hundred percent of the total number of "things" born into the world that subsequently experienced the ratio of deaths shown in the d_x and d'_x columns.

The next step is to recognize the pattern of survivorship. Even though the original data was a record of deaths (d_x column), we must acknowledge that these things once were alive. Any "thing" that lived long enough to die in the second year obviously survived the first year. Similarly, from the d'_x column, if a total of 1,000 "things" finally died, they obviously had to be born first. Therefore, the survival column begins with one hundred percent of the subjects, 1,000, as if they entered the scene at age 0, that is, upon the event of their birth. So, beginning at the top of the fourth column, labeled "Number of Survivors" or l_x, to signify "living at age x," the cohort of 1,000 is entered as if they were born into the world. Reading from the d'_x column it can be seen that 333 deaths occurred *during* this first time period. Therefore, there were 667 which survived to *begin* the *second* time period. *During* this time period, 267 died leaving 400 survivors to *begin* the *next* time period, x_3. The rest of the table is easy to run down. A check on the arithmetic of the survivor column shows that it comes out to zero; that is, the survivors entering the last age group is exactly equal to the number that will die in that time period. Thus, sixty-six survivors beginning this time or age period minus sixty-six deaths *during* the same period equals zero. This also indicates a natural longevity of five years for this cohort. None of them lived any longer!

Appendix B

The Complete Life Table

As described in Chapter IV, there are several sources of data from which life tables can be constructed. Once the basic calculations are made (Table 1A), either from an array of death statistics (d_x column) or from a census (l_x column), the next conventional step is to calculate the probability of death at each particular age. This has been described in Chapter IV as calculating the value for q_x, which is shown in the fifth column of the complete life table, Table 1B.

Calculating Mean Expectation of Life

It remains now only to find e_x, mean expectation of life at birth or at any given age. This is a most significant vital statistic revealing some of the more profound demographic parameters. To calculate e_x, first an auxiliary column, L_x, must be derived from the l_x column. L_x is a symbol for the demographer's gimmick to express

TABLE 1B
Model of a Complete Life Table

x	d_x	d'_x	l_x	q_x	L_x	e_x
1	5	333	1,000	333	833.5	1.83
2	4	267	667	400	533.5	1.48
3	3	201	400	505	299.5	1.16
4	2	133	199	669	132.5	0.84
5	1	66	66	1,000	33.0	0.50

"people years lived" and is used for estimating how much longer a person can hope to live if he has reached some particular age. It is like saying that so many people lived so many years, something like the idea of man-hours of work. The figures in this column are obtained by adding the number of survivors shown in the l_x column for two successive age classes, and dividing by two. Thus, $1,000 + 667 \div 2 = 833.5$, or 833.5 person years, on the average, were lived during the first year of life by the cohort of 1,000 that began life together. In other words, 1,000 began life in the first year, but only 667 finished the full year. So on down through this column, the midpoint of people-years-lived, like so many man-hours of work, is found by averaging each successive pair of survivor groups. For example, L_x at $x = 2$ is

$$\frac{l_{x2} + l_{x3}}{2} = \frac{667 + 400}{2} = 533.5$$

There is a bit of a puzzle in calculating L for the last age class, for of course there isn't any age class after the last one. So, a zero is assumed for an age class beyond the last one because the census or sample did not find anybody alive after this age class. In this example, therefore, sixty-six is averaged with zero; thus, thirty-three is entered as the average number of survivors of this last age to live halfway through this time period ($66 + 0 \div 2 = 33$).

After the L values are obtained for each age class, computations to derive mean expectation of life can be begun. Now for another sublety! Before a mean expectation of life can be calculated for a cohort, the last member must live out his life-span in order to show how many people-years, again like man-hours, that this last member contributed to the overall average life-span of the whole cohort. From this point of view the expectation of further life for a particular age class depends on the whole life history of the cohort to *follow* this particular age class. Therefore, calculations begin from the *bottom* of the L column and work up to each successive age class until, finally, e_x for the newborn cohort is computed.

The formula is really a sort of cookbook procedure which reads

$$e_x \text{ for any age class} = \frac{L_n + \text{all L's up to a given age x}}{l_x}$$

$$n = \text{oldest age class}$$

A few examples will illustrate how simple the process is to compute. From the table, if we should want to know the e_x* for age class 3 (that is, $x = 3$)

$$\frac{L_5 + L_4 + L_3}{l_3} = \frac{33.0 + 132.5 + 299.5}{400} = 1.16 \text{ years}$$

$$e_x, \text{ if } x = 4, \text{ is simply } \frac{33.0 + 132.5}{199} = 0.84 \text{ years}$$

---

* U.S. Bureau of Census usually prints this as e_x^0; the ^{0} means "from birth" and the x means "to this age."

Mean expectation of life at birth for this whole co-hort is therefore the sum of all Ls divided by 1,000, which is 1.83 years. This completes a description of a life table in a very simple model of five age classes. A life table for human beings would contain about one hundred age classes. However, even though the model has only five age classes, it is quite realistic for many wild rodent populations because many of them do not have an ecological longevity beyond five years.

References

Ardrey, Robert. *The Territorial Imperative: A Personal Inquiry into the Animal Origins of Property*. New York: Atheneum Press, 1966.

Aristotle. *Politike*. Translated by H. Rackman in the Loeb Classical Library, New York. Also in U.N. Population Studies No. 17.

Birdsell, J. B. "Some Environmental and Cultural Factors Influencing the Structuring of Australian Aboriginal Populations." In *Human Ecology: Collected Readings*, edited by Jack Bressler. Reading, Mass.: Addison-Wesley, 1969, pp. 51–89.

Borgstrom, Georg. *The Hungry Planet*. New York: Collier, 1967, pp. 7–8.

Brentano, L. "Die Schrecken des uberwiegenden Industiestaats." Berlin, 1902. Cited in U.N. Population Studies No. 17.

Chalmers, T. "On Political Economy in Connexion with the Moral State and Moral Prospects of Society." New York, 1832.

Chapman, R. N. "The Quantitative Analysis of Environmental Factors." Ecology, vol. 9 (1928):111–22.

Erlich, Paul R., and Erlich, Anne H. *Population, Resources, Environment: Issues in Human Ecology*. San Francisco: W. H. Freeman and Company, 1972.

Forrester, Jay W. *World Dynamics*. Cambridge: Wright-Allen Press, 1971.

———. *Urban Dynamics*. Cambridge: MIT Press, 1969.

Ginsberg, Norton S., et al., eds. *Aldine University Atlas*. Chicago: Aldine Publishing Company, 1969.

Huan-Change. "The economic principles of Confucius and his school." Cited in U.N. Population Studies No. 17.

Malthus, T. R. "An Essay on the Principle of Population." London, 1798.

Markham, S. F. *Climate and the Energy of Nations*. New York: Oxford, 1947.

Paludin, Knud. "Contributions to the Breeding Biology of *Larus argentatus*." Copenhagen: Kommission Hos Ejner Munksgaard, 1951.

Patten, S. N. "Essays in Economic Theory." New York, 1924.

Paynter, R. A., "The Fate of Banded Kent Island Herring Gulls." *Bird Banding*, vol. 18 (1947):156–70.

Smith, A. *Inquiry Into the Nature and Causes of the Wealth of Nations* (1776). New York: Modern Library edition, 1937.

Steinberg, S. H., ed. *The Statesman's Yearbook*. New York: St. Martin's Press, 1971.

Sumner, W. Graham, and Keller, A. G. *The Science of Society*. New Haven, 1927.

Toynbee, A. "A Study of History." London, 1939.

United Nations Population Studies No. 17. *The Determinants and Consequences of Population Trends*. New York, 1953.

United Nations Population Studies No. 29. *Multilingual Demographic Dictionary*. New York, 1958. In three languages: English, Spanish, French, separately.

Suggested Readings

Bates, Marston. *Man in Nature.* Englewood Cliffs, N. J.: Prentice-Hall, Inc., 1965.

Boughey, Arthur S. *Ecology of Populations.* New York: Macmillan Company, 1968.

Kormondy, Edward J. *Concepts of Ecology.* Englewood Cliffs, N. J.: Prentice-Hall, Inc., 1969.

McGaugh, Maurice E. *Geography of Population and Settlement.* Dubuque, Iowa: Wm. C. Brown Company, 1970.

Meadows, Donella H., *et al. The Limits to Growth.* New York: Universe Books, 1972.

Wilson, Edward O., and Bossert, William H. *A Primer of Population Biology.* Stamford, Connecticut: Sinauer Associates, Inc., 1971.